# Palgrave Studies in Media and Environmental Communication

Series Editors
Anders Hansen
School of Media, Communication and Sociology
University of Leicester
Leicester, UK

Steve Depoe
Mcmicken College of Arts and Sciences
University of Cincinnati
Cincinnati, OH, USA

Drawing on both leading and emerging scholars of environmental communication, the Palgrave Studies in Media and Environmental Communication Series features books on the key roles of media and communication processes in relation to a broad range of global as well as national/local environmental issues, crises and disasters. Characteristic of the cross-disciplinary nature of environmental communication, the books showcase a broad variety of theories, methods and perspectives for the study of media and communication processes regarding the environment. Common to these is the endeavour to describe, analyse, understand and explain the centrality of media and communication processes to public and political action on the environment.

More information about this series at
https://link.springer.com/bookseries/14612

Benjamin J. Abraham

# Digital Games After Climate Change

palgrave
macmillan

Benjamin J. Abraham
University of Technology Sydney
Sydney, NSW, Australia

ISSN 2634-6451          ISSN 2634-646X   (electronic)
Palgrave Studies in Media and Environmental Communication
ISBN 978-3-030-91704-3          ISBN 978-3-030-91705-0   (eBook)
https://doi.org/10.1007/978-3-030-91705-0

This Palgrave Macmillan imprint is published by the registered company Springer Nature
Switzerland AG.
The registered company address is: Gewerbestrasse 11, 6330 Cham, Switzerland

*To Samantha*

# ACKNOWLEDGMENTS

This book project started with the realization that if I was going to take climate change seriously I needed to do something about it. This book is that attempt to do something, and I hope it guides and encourages others to do something as well. From the first presentation in 2014 at the Digital Games Research Association (DiGRA) Australia conference of the very initial ideas that would eventually turn into this book, they have been shaped and guided by so much help and encouragement from friends, family and colleagues that there are almost too many to thank. After such a long and drawn out writing process, it also has the tendency to becomes a bit of a blur, increasing the risk of omitting thanks to people who helped make this book happen. My deepest thanks go out to everyone who has helped along the way, whether with feedback and suggestions, or even simply an encouraging word.

I want to thank in particular my colleagues in the School of Communication and the Climate Justice Research Centre at the University of Technology, Sydney, especially my colleagues in DSM. Thanks, in particular to the legendary trio of Liz Humphrys, Sarah Atfield and James Meese who always found time to listen to me over the years and without who's support I would not have survived the tough years. Thanks to Darshana Jayemanne and William Huber of Abertay University's games program for hosting me during my visit in 2019, for giving me the opportunity to test the ideas in this book with their excellent students, and for taking me for the most memorable yuzu shaved ice dessert in Kyoto after DiGRA. I have to thank the many academic mentors I've had the privilege of having over the years, from whom I have learned so much about the academy, research, writing and being part of an intellectual community: my PhD supervisor Maria Angel, my career guru Tom Apperley, my intellectual idol Andrew Murphie, and many, many, many others. Without the extremely generous support of the UTS Science Laboratory, particularly Dr. Ronald Shimmon and Dr. Dayanne Mozaner Bordin, the analysis of the PS4 APU and the whole of Chap. 7 would have been completely impossible. I am deeply in their debt for taking on the project with the pathetic "budget" I had. I also owe much to Christian McCrea, both for encouragement and guidance, and for providing the phrase 'periodic table of torture' to describe the results of this same analysis. My deepest thanks as well to all the game developers who responded to my survey in early 2020, providing energy usage figures and power bills that enabled Chap. 4 to be as detailed as it is.

This book has benefited immensely from the expertise and attention of various readers of drafts and chapters: Hugo Bille, Julius Adamson, John Groot, Mark Videon, Eric Zimmerman, Jackson Ryan and others—as well as the encouraging environment and occasional input from members of the IGDA Climate Special Interest Group, the journalist and game industry groups I have spoken to about the project and shared initial findings with. Without the constant, unconditional support of the most tight-knit gaming community I have ever been involved with—The Shoot Fam—I would have gone crazy long before finishing: may all your drops be god rolls. Similarly, my closest friends 'The Dads'—Terry Burdak, Brendon Keugh, and Daniel Golding—all my gratitude and love to you three, I could not ask for better mates. My parents, Alison and Craig, who have always encouraged me to pursue my weird, niche, and more out-there projects, and unconditionally supported me through all my questionable career decisions: I owe you both more than words can convey. This book is in no small part a reflection on your patience and generosity, as wonderful parents as could ever be hoped for. To my brother Nick, who was right there with me for so many formative gaming experiences, playing *Halo 2* co-op over sweaty summers holidays. And finally, to my partner Sam, who has put up with far more grumpiness and exasperation from me than she should ever have to, as I tore my hair out, swore and shouted and generally procrastinated finishing this or that—this book is dedicated to you. Love you Sammy.

I hope this book acts as a catalyst for change and stands as a signpost for how to make that change happen. I hope it serves as a testament to the importance of climate action right now in every part of both our work and leisure, and the transformation of the digital games industry which I know so many of us care such a great deal about. Any errors or mistakes remaining are mine and mine alone.

# CONTENTS

# LIST OF FIGURES

# CHAPTER 1

# Why Games and Climate Change?

Games and heat are inextricably linked. Growing up in Australia, playing games during summer holidays meant periods of grueling physical endurance punctuated by retreats to cooler parts of the house. As a young teenager, my childhood room was on the top floor of my parents' split level and had a large single-glazed window and thinly insulated roof. These materials offered poor protection from the blazing sun and ambient air temperature outside, which often built up precipitously over the long, unrelenting Australian summertime. Temperatures both outside and in often reached over 40C. My room would remain at this temperature hours after the outside had cooled down, the environmental heat compounded with the additional burden of waste heat from my gaming PC or chunky CRT television.

When I think about the future of digital games in the context of our warming planet, I often return to this experience, and the profoundly pragmatic decisions it forced upon me: sometimes it was simply too hot to play games, too hot to sleep. Often I entered into a calculus about moving to a cooler part of the house, or whether I could afford to play videogames and further heat the room (I almost always did, and suffered the consequences). The dilemma of what to do when it becomes too hot to play games, and the questions it raises about where to go, or what to do about the waste byproducts of games (like heat) informs the perspective this

© The Author(s), under exclusive license to Springer Nature
Switzerland AG 2022
B. J. Abraham, *Digital Games After Climate Change*, Palgrave
Studies in Media and Environmental Communication,
https://doi.org/10.1007/978-3-030-91705-0_1

book takes on the increasingly pressing questions and multiple, intercon-nected dilemmas facing the global games industry within the context of our rapid heating world.

It has been all too easy, at least until very recently, to treat climate change as something 'coming', as a problem for tomorrow rather than today. However, recent events have started to erode the sense that there may be some time left, some distance between 'now' and the 'then' in which our climate changes. Climate change has arrived, both as a gradual ratcheting up of a background risk and as new normal increasingly punc-tuated by extreme events; from the devastating El Niño amplified drought that precipitated the catastrophic Australian summer bushfires of 2019–20 which burned the length of the continent; to the record breaking cyclone season of 2020 which saw tropical cyclones of unprecedented strength like Super Typhoon Goni, that devastated the Philippines; to the most damag-ing monsoon flood season in Asia, that caused $32bn damage in China alone, killing 278 people (Nuccittelli & Masters, 2020). The growing ferocity and impact of extreme weather events begins to threaten to dis-rupt modern ways of living which depend on uninterrupted global supply chains, predictable patterns of energy demand, and weather conditions that are conducive to workers being able to simply go to work every day. All of the threats to these systems will eventually make their way home to the various spheres in which we make and play digital games. If carbon emissions continue to rise unchecked and the planet continues its precipi-tous trajectory, then serious and significant global shocks of an unpredict-able nature and scale are, even according to the most conservative scientific voices, going to threaten everything about how we live our lives today (Flannery, 2005; IPCC, 2018; Rockström et al., 2009; Dull et al., 2010; Atwood, 2015; Parker, 2017). These threats are, of course, not evenly distributed nor are they faced uniformly across the globe. There are deeply political, often deeply unjust, consequences to the fact that both the great-est number of people and those who are the most exposed to climate threats tend to be located in poorer, less developed nations. It is partly for this reason that one of Wainwright & Mann's (2018) political futures out-lined in their book *Climate Leviathan* involves the rise of a 'Climate Maoism' out of an Asian experience of being on the front-lines of climate disasters, and an unwillingness to bear the brunt of Western nations' fossil fuel profligacy. An outlandish possibility to some, perhaps, but certainly not beyond the realms of possibility. What sort of challenges and conse-quences would the games industry face in a world dominated by a populist climate dictator devoted to harnessing the frustrations and channeling the demands of the masses worst affected by climate disasters? What happens when the current already-shaky neoliberal political order falls, and proves inadequate to the task at hand? In a future facing real resource and carbon constraints, how highly would we priorities the making and playing games? What if we had to choose whether to play games or run the air conditioning? Are we ready for that?

In a context of "everything change," as Margaret Atwood (2015) describes the climate crisis, games will increasingly be seen as integrated with the rest of life. So while it might seem trivial or perverse depending on where one is situated to be concerned with the fate of a leisure industry in the face of arguably more serious disruptions, this seems to me all the more reason to start the process *now* of thinking about what climate change means for games, and what they can be done about it before being forced to act by events outside our control. Bringing the climate crisis 'home' to all of our lives, all of our workplaces, all of our hobbies, is the necessary first step in acting to reverse climate action. We need to start to understand what those actions could be, or will need to be, as well as what constraints we are likely to face. The eight chapters that form this book raise some arguments that, no doubt, will be contentious, and are by no means the final say—things are changing rapidly, and the research and findings contained in this book may well shift as things like the renewable energy transition gathers speed, as the climate crisis continues, and hopefully as more and more people and organizations get on board with serious, sustained action. Beginning with more traditional game studies theory and analysis (Chap. 2), including a close reading of games in the 'survival-crafting' genre (Chap. 3), the book moves beyond these to focus on how games get made, and the associated climate and environmental costs. It is a picture that some readers may be uncomfortable with, and draws on literature that does not typically show up in either academic or industry books on videogames—from corporate reporting documents to electronics handbooks and scientific research on mining processes and emissions. I try and make these discussions as accessible as possible to the general reader but some complexity is unavoidable given the subject matter. One whole chapter (Chap. 7) is devoted to the molecular makeup of the electronics in the PS4, tracing a speculative 'periodic table of torture' that entails a diffuse and tangled network of harms. These emerge from the results of an inductively coupled plasma mass spectrometry (ICP MS) test, a procedure that analyzes the elemental makeup of the otherwise very ordinary consumer electronic device that is the *PlayStation 4*—already rapidly approaching obsolescence—finding a host of exotic substances. The analysis allows for the identification and subsequent investigation of the most basic elemental components of the console's main chip, where they have come from, and what sorts of work and emissions are entailed in producing them.

Whatever our climate future holds, change is coming to games, and it is coming faster than we might expect. Based on what we know about planetary bio-physical limits, as well as the political and economic constraints we face, it seems quite likely that we may, much sooner than we would like, be faced with some stark choices—choices as an industry, as researchers and scholars, as designers and players. Choices like whether to launch a new console platform with new and improved specifications, or making do with existing hardware, maintaining it for longer, squeezing more juice out of it through other means. Choices between focusing our research efforts and our advocacy on things that might benefit a modest number of players or that might benefit everyone through concrete and measurable reductions in actual emissions. It is clear that a conflict is brewing between the way the existing digital games industry is economically and politically organized and our goals for a livable planet. Facets of this conflict are becoming apparent as well; conflicts between the regimes of player engagement the industry creates through game design, marketing and genre conventions, and the need to decarbonize and reduce environmental footprints of its now massive player base; conflicts between the reliance on a 'new' hardware generation every few years, unrepairable and un-upgradeable devices, and the urgent need to sustainably use precious, finite resources; conflicts between the need for large corporations to produce a sufficient rate of return for investors, and the critical importance of acting ethically and responsibly in the world.

Limiting the effects of catastrophic climate change will hopefully (and if it has not already) lose its status as 'optional' to political electorates, as the increasing evidence of our worsening predicament swirl around us almost daily. Climate issues are poised to become a central organizing and determining feature of all of contemporary life, work, and leisure to a degree we have not seen since perhaps the Second World War. What this means for those of us in the field of games studies, what it means for developers and others who work in and around the industry itself, from streamers to journalists, to the players that exist at the end of the huge chains of logistical networks and infrastructures that form the machinery of the modern games industry—that is what this book sets out to just begin to describe. It is a huge task, and one that I cannot hope to complete. It is the aim of this book to 'open the door' so that others might take up these challenges, take these necessarily broad sketches and flesh them out in much greater detail. The scope of this task is reflected in the length of the book (longer than I'd like). It is my hope that in spite of this, readers can find the chapters that provide most what they are after. For scholars inter-ested in debates about games effect on players, Chaps. 2 and 3. For mem-bers of the game development community—Chaps. 4, 5, and 6 in particular may help provide a concrete sense of the scope of the industry's emissions and why it is so important to do something about them. For scholars interested in the materiality of media, Chap. 7 may hopefully offer new details into the actual *stuff* and workings of computation, and provide new avenues of investigation for mapping and analyzing high tech devices that rely on the same sorts of computational machinery as games hardware.

While most of the chapters can be read on their own, there is also a broader argument that stretches across the arc of the book about the necessity of concrete action on the emissions intensity of the games indus-try today and why (and how) that must be our top priority. It is my firm belief that an industry-wide commitment to carbon neutrality is utterly essential. Anything less than that is simply insufficient. Before we get to that discussion, however, it is important to get a sense of how and why games are not innocent when it comes to the climate crisis—where their industrial and material outputs are plugged into the same global systems using up the earth. We also need to understand and locate the causes of the climate crisis, before we can begin our search for adequate and appro-priate remedies.

## COSTLY DIVERSIONS

The games that we have been playing, for all the innocent pleasures they may bring, are profoundly entangled with the global processes that are fueling and deepening the climate crisis. Games are played on hardware that is energy intense and generative of significant ecological harms, both at the time of use and over a device's lifecycle. Made from minerals dug out of the ground using fossil fuels, packaged in plastics derived from pet-rochemicals, designed and assembled often under intense labour condi-tions, and finally shipped around the world as part of global supply chains producing value for shareholders. Each of these are processes that are increasingly well-documented and understood, including for their role in our current crisis. This knowledge has yet to have much purchase on either the strategic outlook of the industry itself, on the nature of the games that are made and sold, or on the consumer demand that businesses are ultimately responding to. Games as a sector require vast amounts of electrical energy while releasing noticeable amounts of thermal energy into the sometimes already quite heated spaces in which we play (as any old school LAN party attendee can attest). When we are done with our discs and devices, when they break, they frequently return to the ground to leech heavy metals and other toxic mixtures of both rare and common elementary materials back into the heterogenous earth that is called 'landfill'. Recycling of electronic devices like mobile phones and LCD TVs barely begins to make a dent in this process, and the games industry has so far largely escaped significant political or consumer activism to force the issue of the end-of-life disposal or recycling of computer hardware.

None of this can be separated from the dominant political economic system and its global effects that are felt disproportionately and unevenly in the global south. Whether the game industry can be made to acknowledge these issues, and how it responds to these challenges will be the measure of its ability to occupy a legitimate place in a world with a changed climate. It also still remains to be seen whether the games industry is capable of the depth and maturity of leadership that has thus far often eluded it. Thinking along climate and energy intensity lines, I hope this book shows, can also be an incredibly fruitful and productive exercise for an industry which has long been (or, at the very least long been viewed as being) mired in a kind of perpetual adolescence: an industry of toys for boys. There are encouraging signs that this is changing (Golding & Van Deventer, 2016; Chess, 2017; Ruberg, 2020). Thinking through the implications for climate action within the games industry could be one more front on which to pursue real advances in the industry, to pressure for yet more of the necessary 'growing up' which still remains to be done.

Gaming is still, by and large, a leisure activity—and presently it is a relatively carbon intensive one. Chapters 4, 5, and 6 attempt to put as concrete a figure as possible on exactly *how* intense, in order to move forward the conversation towards the essential work of abatement and mitigation. To know what sort of environment games and game players might face in the future—both in the sense of our physical-climactic environment and the political economic environment—it is necessary to understand the nature of the unfolding crisis as well.

## CONCEPTUALIZING THE CRISIS

Anthropogenic (human driven) climate change can be understood in any of a number of different ways. Most obviously, it is an ongoing process of increasing $CO_2$ levels in the atmosphere and dissolved in the oceans; it is the 'greenhouse effect' of trapping more solar radiation in the atmosphere, triggering rising temperatures and sea levels; it is an increase in the statistical 'risk' of extreme climate events from this greenhouse effect; it is the consequence of a loss of carbon sinks like rainforests due to land clearing for farming and development, further accentuating the effects of already high emissions levels, and so on. Many of these formulations, however, merely describe the symptoms rather than the cause of the crisis itself. One can diagnose the problem of climate change in a way such that the analysis finishes at its proximate cause—the release of carbon into the atmosphere—without ever really examining why it is occurring. One can even do so quite easily, and without recourse to a particular politics or entailing any given action, as many climate scientists wish to do. As Australian climate scientist Tim Flannery (2005) points out, it is not even that difficult actually to pull carbon out of the atmosphere, but it is needed at such a scale that it would be a truly global effort. Flannery's (2017) favored solution (one of many) is mid-ocean kelp farming, deploying fast growing seaweed to draw down significant global emissions. Another widely reported paper analyzed the financial cost of planting and managing a massive international program of reforestation across the globe (Bastin et al., 2019). The trouble with both ideas, as good and perhaps even necessary as they are, is that they don't do anything to address the underlying problem. A similar narrowness of thinking leads to other inadequate responses such as engineering solutions like carbon capture and storage, or the belief that carbon simply needs to be 'priced in' to markets, both of which do little if anything to change the underlying structural—which is to say political and economic—causes (Bryant, 2019; Moore, 2015; Wainwright & Mann, 2018).

In another sense though, this myopic view remains the somewhat valid double of a larger, more encompassing critique—climate change is actually only an existential problem for us and other forms of life that evolved or existed during the Holocene (the last ten thousand of so years). The planet's alleged crossing of a geological threshold into the deadly Anthropocene (Zalasiewicz et al. 2017), even the planet's sixth major 'extinction event' will not cause the universe, or even life on the planet to cease to exist. It might well persist in some other form, as many are realizing. It is not immodest to recognize that the world (changed as it is) will almost certainly 'go on without us', despite it being in a more rudimentary, even an unrecognizable form. And even if it does not, it would be an act of incredible hubris to think that 'us' continuing (even extending 'us' to include all life on earth) is all that matters. This somewhat nihilistic perspective has a few things going for it, and has found a number of advocates—like Chris D. Thomas (2017) whose book *Inheritors of the Earth* looks not for the unfolding of a grand tragedy but at the surprising success stories of those species which are thriving in human-altered environments, whether in spite of or because of us. And to a degree, parts of this argument are successful I think—but as a whole it requires us to be willing to let go of a great many things currently important or valued, from the way we live our lives, and indeed the way many other species live theirs. Crucially, however, many don't even get a say in it. But then again, neither do we, as the climate crisis reveals the profound democratic deficit in the global political order. We must also acknowledge the unmitigated, prolonged human suffering on a global scale that we would be unleashing should we accept a 'business as usual' approach—perhaps the kind of suffering not seen since the black death in Europe, or the human-exacerbated El Niño famines that Mike Davis (2002) describes in *Late Victorian Holocausts*. We must acknowledge that it represents a repetition and intensification of the experience the transatlantic slave trade and the "black and brown death [that] that is the precondition of every Anthropocene origin story" as Yusoff (2018) describes. The last time that the world experienced such a profound climate shift, during the seventeenth century, virtually no parts of the globe remained unscathed. Nations and empires experienced some of the greatest political upheavals in modern history, as Geoffrey Parker (2017) chronicles in *Global Crisis: War, Climate Change and Catastrophe in the Seventeenth Century*. Most unfairly of all, this suffering will be—is already being—borne most often and most heavily by those who have done the least to contribute to it, a terrible injustice (Methmann & Oels, 2015). Thus, this book rejects the option of doing nothing, letting things take their course, and refuses to countenance human extinction—preferring instead to imagine the extinction and replacement of the current fossil fuel economy.

Climate change already presents a challenge to quotidian human life today by undermining our sense of a future, of something to look forward to. David Collings (2014) argues that the effects of climate change are already being felt in our mental landscape, arguing that it has stolen the future:

> Climate change does not just melt the ice caps and glaciers; it melts the narrative in which we still participate, the purpose of the present day. In this sense, too, we are already living in the ruins of the future. (Collings, 2014: 116)

Perhaps because of the trauma involved in confronting what we are doing to the planet, Collings (2014) argues climate action has been rather successfully co-opted by "green consumerism" and marketed for feel-good lifestyles that do nothing to address the underlying causes. Climate change, he argues,

> is nothing less than an assault on who we think we are: it exposes the fact that the economies of the developed world are founded on a lie, that our way of life takes for granted the eventual destruction of the Earth, and that persisting with it makes us complicit in a great crime. (Collings, 2014: 17)

Similar arguments have been made by others, including Naomi Klein (2015) in her book *This Changes Everything*. Klein (2015) argues that we have actually undersold both the severity of the problem and the amount of work required to solve it, and that by thinking we can leave climate mitigation up to the market, we have been lulled into a way of thinking that pretends that it is sufficient to continue on with our lives and lifestyles as they are. This, in concert with scientists and experts who wish to avoid seeming 'alarmist' in their predictions, has seen climate action treated as only a lifestyle choice, the preside of the wealthy and well-to-do urban types, and the assumption that, whatever else happens, capitalism will surely at least ensure its own survival (a dangerously untested assumption). In reality it is becoming apparent that the best and most developed models of the changing climate all suggest a radically different prognosis, and that a much greater mobilization across all sectors of life is required to decarbonize the world and shift our economies. I suggest starting this process with what we know, and with what is immediately before us—in our case, with games and the context they are made in, the people who make them, and the industry that makes them possible.

If climate change demands we understand it beyond just proximate causes, how do we make sense of it in a way that doesn't lead to the paralysis or despair for its scale and magnitude? How are we to grapple with something so seemingly massive and expansive? Timothy Morton (2010) turns this problem on its head, drawing a kind of insight from how our thinking already responds to climate change. Morton (2010: 1) offers the following:

> The ecological crisis we face is so obvious that it becomes easy – for some, strangely or frighteningly easy – to join the dots and see that everything is interconnected. This is the ecological thought. And the more we consider it, the more our world opens up.

The challenge thinking ecologically presents us with is as intimidating as it is alluring, and it offers a host of new and unexpected connections and opportunities. A commitment to attempt to think ecologically informs the following chapters, representing the closest thing to a 'method' for this book, explaining why I refuse to settle into any one disciplinary stream. The challenges before us will be as much about solving physics challenges regarding efficient use of energy, and complexity problems around where best to direct our scarce resources, as much as questions of how and where (even if) players can be convinced to act 'greener' through the games they play.

One consequence of Morton's insights is that once we begin to think ecologically it becomes incredibly difficult to stop. Previously separate issues and distant locations, are suddenly revealed to be much more intimately connected than we thought. If we are curious, if we want to get to the bottom of things we simply cannot stop ourselves from thinking ecologically. This book is an attempt to apply this thought to the sphere of games, a sphere which has tended to be approached primarily through the lens of games-as-played, existing often within a distinct realm of 'culture'—one which might even be said to replicate a problematic human/nature binary (Keogh, 2018). Less often, though increasingly so, games researchers have begun to undertake more ambitious analyses of games as material objects, and as a product of industrial and political processes ranging from mineral extraction to labour exploitation (Dyer-Witheford & de Peuter, 2009; Montfort & Bogost, 2009; Taffel, 2015). The 'material turn' of game studies as described by Apperley and Jayemanne (2012) is still difficult to reconcile with the fleeting, ephemeral state of the experience or moment of play, but this is changing. The approach here is not quite a historical-materialist analysis, but it remains deeply sympathetic to such approaches, many of which helpfully force us to encounter a polity of humans and nonhumans, echoing the work of scholars of non-human agency like Jane Bennett (2010), Bruno Latour (2005), Michel Callon (1986) and many others.

There are other big theoretical questions that need to be addressed in our responses as well. Jason W. Moore (2015) is one of a growing number of thinkers who have begun to do the essential work of understanding the political economic origins of climate change. As noted earlier, it's entirely possible to stop one's analysis at the physical causes of climate change (if we are not thinking ecologically), leaving us without a clear sense of how to respond, or what the systemic causes are of the increase in emissions. Moore's (2015) work, which combines both history and political economy, reaches conclusions that defy the typical disciplinary boundaries of either field. He argues that a major contributor and guarantor of our current situation has been Western philosophical and conceptual orientations towards the very question of what it means to be human, and what this conception implies about Nature—with an emphatic capital N. This idea of Nature is 'out there', it is 'wilderness', an untouched or untamed space free from human interference and always counterposed to the realm of culture, industry, science, knowledge. Crucially, in this state it exists to be harnessed and drawn upon as part of the hundreds of years long process that has been the advent and development of capitalism. For Moore (2015), much of our current thinking around environmentalism and sustainability is inflected with what he argues is a type of 'green arithmetic'. A form of Cartesian dualism, it assumes that if we just 'add nature' back to whatever our current ways of thinking are that we will achieve a sufficient understanding:

> The Cartesian narrative unfolds like this. Capitalism – or if one prefers, modernity or industrial civilization – emerged out of Nature. It drew wealth from Nature. It disrupted, degraded, or defiled Nature. And now, or sometime very soon, Nature will extract its revenge. Catastrophe is coming. Collapse is on the horizon. (Moore, 2015: 5)

Moore is of course not the first to critique either this type of Cartesianism (e.g. Grosz, 1994) or even necessarily the green arithmetic perspective: thinkers as diverse as Bruno Latour (1993, 2005), Donna Haraway

(1991), Val Plumwood (1993, 2009) and others in a range of related fields have provided their own critiques of the same flaw in Western thinking. Moore's analysis however clarifies our understanding that 'capitalism is not an economic system; it is not a social system; it is a way of organizing nature' (Moore, 2015: 2). By nature, however, he does not just mean the 'out there' of the natural world, rather the entire world that opens up before us once we start thinking ecologically. In this view, there is nothing either natural or unnatural about the rare earth minerals present in a game console, or about a teenager sitting at a computer in an air-conditioned room playing games. This version of nature includes all that exists, humans and all the things we fashion. To be human means to be human-in-nature—there is no separation, and there are not some of us more or less 'in-nature' than others. One distinct advantage of this approach is that it does not predispose us to a hatred of technology, or to endorse some form of climate primitivism where we have to give up everything we currently enjoy. It does not force us to disavow the pleasures of digital games and technologies, but instead it simply shows that our technological achievements are constrained by capitalism's organizing imperatives, and that they could be harnessed and directed for entirely different ends. This is not the way of climate asceticism, of visions of shrinking ourselves and our horizons, of always 'doing less with less'. It has more in common with the expansive dream of socialist utopias, or the "Red Plenty" that Francis Spufford (2010) so effectively dramatizes in his book of the same name, more about balancing ecological systems, doing more with what we already have, and of closed loops of creative repurposing.

Importantly from a climate justice perspective, Moore (2015) notes that the view of humans as separate from nature 'is directly implicated in the colossal violence, inequality, and oppression of the modern world; and that the view of Nature as external is a fundamental condition of capital accumulation' (Moore, 2015: 2). There are practical consequences to this shift in perspective:

> The most elementary forms of differentiation...unfold as bundles of human and extra-human natures, interweaving biophysical and symbolic natures at every scale. The relations of class, race, and gender unfold through the oikeios; they are irreducible to the aggregation of their so-called social and ecological dimensions. (Moore, 2015: 9)

In other words, there is always more than whatever we get when we add 'humans' and 'nature' together—there is an overflow, an excess, a world that escapes and expands, unable to be captured by this operation. Moore (2015: 5) concludes that what is needed is 'a new language – one that comprehends the irreducibly dialectical relation between human and extra-human natures in the web of life – [which] has yet to emerge.'

Given this discussion so far has largely conceptualized the problem underpinning climate change as an issue with Western thought and Western dualism, I have really only treated the problem with more Western thought. This might be a case of trying to use the master's tools to dismantle the master's house (Lorde, 1984)—alternatives exist outside the western canon as well. Much non-Western and indigenous thought has had no trouble at all conceptualizing the sympathetic and reciprocal relationship between people and planet, avoiding these sorts of problematic dualisms. Linda Tuhiwai Smith (1999) in her landmark work *Decolonizing Methodologies* describes the way indigenous thought often presupposes a humanity-in-nature perspective, akin to what Moore (2015) thinks is needed. Smith (1999: 105) notes that, beyond simply conceptual contributions, 'indigenous peoples offer genuine alternatives to the current dominant form of development. Indigenous peoples have philosophies which connect humans to the environment and to each other and which generate principles for living a life which is sustainable, respectful and possible.'

In an Australian context, encouraging work done by both Indigenous and non-Indigenous scholars has helped to raise the profile of Aboriginal and Torres Strait Islander knowledge and understanding of deeply practical questions of living with our environments. Bruce Pascoe's (2018) *Dark Emu*, and as Bill Gammage's (2011) *The Biggest Estate on Earth* both do important work reconstructing the early evidence for the knowledge held by First Australians in relation to the distinctive environment, and in particular its active fashioning and sustainable management that encouraged the flourishing of complex ecologies of life. Deborah Bird Rose (1996: 3–4) worked with and learned from Indigenous Australians for a number of years, and notes that 'Aboriginal people have developed a system of knowledge and a way of managing the continent that is quite different from the ways that European-derived cultures manage knowledge and land.' Widely accepted amongst the diverse linguistic and cultural groups that make up Australia's Aboriginal and Torres Strait Islander nations is a shared respect and stewardship of 'country', a unique sense that isn't often captured by the non-Indigenous meaning of the term: 'country, to use the philosopher's term, is a nourishing terrain. Country is a place that gives and receives life. Not just imagined or represented, it is lived in and lived with' (Rose, 1996: 7). The Indigenous Australian attitude to the environment manages to acknowledge it as active, possessing a vitality or agency that is often absent from Western approaches—approaches that too often treat it as a static resource to be drawn upon, a place for soaking up 'externalities', with the atmosphere and the oceans ready to be used up by industrial emissions as private profit extracts from these living systems, from these nourishing terrains, more than they can sustain. According to Rose (1996: 7) 'country is a living entity with a yesterday, today and tomorrow, with a consciousness, and a will toward life.' Country here becomes,

> Ideally...synonymous with life. And life, for Aboriginal people, needs no justification. Just as no justification is required to hunt and kill in order to support one's own life, so there is no justification required in asserting that other living things also want to live, and have the right to live their own lives. It follows that other species, as well as humans, have the right to the conditions which enable their lives to continue through time: minimally to the waters and foods on which they depend, and to the sanctuaries in which they cannot be hunted or gathered or harmed in any way. (Rose, 1996: 10)

Aboriginal and Torres Strait Islander cultures are some of the oldest on the planet, and they reflect a deep history of stewardship, care, and respect for that what we would abstractly consider 'the environment' but which covers a lot more than that phrase can capture. It seems a deeply wise position as well—taking care of 'country' is as much an expression of care for oneself. Rose (1996: 10) explains that 'the interdependence of all life within country constitutes a hard but essential lesson – those who destroy their country ultimately destroy themselves.' This is the same lesson Western philosophy is being forced to confront today across the globe through increasing encounters with planetary limits.

What does it mean, then, to take Aboriginal and Torres Strait Islander knowledge in this domain seriously? Firstly, it means adopting a more modest relation to Western scientific knowledge, and an awareness of its implication and imbrication with colonization, and colonialist thought and action (Smith, 1999). It is not a coincidence that it is principally the West that has been the chief contributor to the carbon emissions and the capitalist development cycle that is the principle driver behind climate change. The Marxist feminist research of Silvia Federici (2004) in *Caliban and the Witch* has gone to great lengths to uncover the role of certain 'demystifying' applications of scientific knowledge, particularly in propagating early physical-mechanical understandings of the human body and in the eradication of belief in magic. She argues this was a necessary first step in the development of capitalism, with the vitality and agency of the natural world presenting a threat to the new social order:

> Eradicating these practices was a necessary condition for the capitalist rationalization of work, since magic appeared as an illicit form of power and an instrument to obtain what one wanted without work, that is, a refusal of work in action. 'Magic kills industry,' lamented Francis Bacon, admitting that nothing repelled him so much as the assumption that one could obtain results with a few idle expedients, rather than with the sweat of one's brow. (Federici, 2010: 141–2)

There are implications, invariably, for our responses to climate change and the extent to which we admit that scientific knowledge is applicable to the crisis. I have argued and will continue to do so throughout this book, that more than simple science-based approaches are required. A shift in thinking and acting, and an appreciation for the human dimensions to problems is essential. A sensitive awareness to human values—in their full diversity and regional specificities—will be needed in any serious climate change activist program, as Ketan Joshi (2020) has shown so effectively in his discussion of community involvement in the planning for new wind power in Australia.

Secondly, accepting indigenous people's knowledge suggests another form of modesty, in reflection of the great and terrible powers we have already exercised as a species. We ought to recognize that the responsibility we have for caring for 'country' has now expanded massively in scope and scale—as Rose (1996: 49) points out, 'relationships between people and their country are intense, intimate, full of responsibilities, and, when all is well, friendly. It is a kinship relationship, and like relations among kin, there are obligations of nurturance. People and country take care of each other.' When we have ceased to take care of country, of each other, even of the planet it should prompt us to reconsider and reconfigure our disciplines, our frames for thinking, our worldviews.

There is a great more that could be taken even from just this brief account, but it serves to point towards a general observation about the importance of non-Western thought for addressing the current critical situation. Other fields have more to add but will not be given an extensive treatment here, particularly the work of feminist ecologists like Val Plumwood (1993), who have similarly been working for decades on describing and untangling the problems with Western dualist thought and its sanctioning of ecologically destructive, even self-destructive actions. The approach I am outlining here makes up the theoretical approach of this book, adopting a permissive application of the ecological thought in the approach to describing the problems, networks and entanglements of a range of issues across a range of scales and terrain, and the ethical responsibility to care that we all share.

## OUTLINE OF CHAPTERS

Given the pressing urgency of the situation we face, Chap. 2 begins with the question that I suspect many readers will have already begun to ask themselves—how can games help save the world? Interest in games that do something to advance efforts at bringing the climate back under control has grown over the past decade or so, turning up more and more, even in mainstream games. Games like *Civilization VI*'s "Rising Storm" expansion, which adds climate change features to its end game, and simulation and educational games like *Fate of the World* that put players in charge of a global force tasked with fixing the crisis. In Chap. 2 I examine the literature on the application of games to persuasive or pedagogical ends. It has been suggested that games have utility for all sorts of environmentally friendly goals, from placing players in scenarios of resource scarcity to encouraging greener consumer choices. An issue I take with much of this literature, however, is its lack of attendance to the complexity of players themselves, and the lack of certainty around their willingness to participate in our designed goals for them. In particular, much of this literature does not sufficiently consider the barriers presented by the mystifying effects of ideology, especially important for issues of climate change. The latest research into climate communication stresses the need for an 'interaction' model of climate communication that focuses on tailoring to the particular processes of meaning-making of the receiver. On the surface this might seem well aligned with games and their oft-emphasized responsiveness via interactivity; however, I see significant shortcomings and challenges. This is due in part to the variability of these processes of meaning-making, the deeply personal and individual nature of ideological beliefs, and the conflict it presents with mass-market imperatives. If games are to make interventions on the scale hoped for, then it seems highly unlikely they can be made sufficiently adaptable to the wide variety of players. In response to this, the chapter ends by turning to the military simulation sandbox game *ARMA 3*, finding in it an alternative route to perhaps coax players into accepting positive visions of climate change via its aesthetic. I conclude by considering what players really even *need* from games, and argue that a better approach than attempting to change the hearts and minds of willing (or unwilling) players is to simply and directly change the world around them, and for the better. In this way we can render the resistances and barriers to climate change acceptance nearly redundant.

If games for climate persuasion are not going to save the world, then the next question becomes what does it take for a game to be considered truly ecological? What makes a truly green game? Beginning with an examination of the ecological resonances within the terms that suffuse gaming discourse—terms like 'environmental storytelling' and even 'emergence'—I consider the way these make claims about correspondence to a natural world or natural behaviors. The games which seem to most embody these ideas are games in the survival-crafting genre, games that are often considered by those that advocate their interventionist or persuasive powers as the most useful for teaching various scientific, environmental or ecological concepts. Titles in the same lineage as *Minecraft* (2011) are often celebrated for appearing to reflect the logics of ecosystems in and through their emergent properties, their systemic nature, or reflection of real-world complexity. Conducting a close reading of several survival crafting games, I find them to be quite limited in their ecological capacity. Much of the main determinants of the dynamics of these games, I argue, are instead closer to *economic dynamics*, embedding teleological notions like the ladder of technological progress, and capitalist logics of accumulation. Instead, I argue that a better, and more productive approach to the truly ecological game involves an altogether different focus, specifically on the game's own existence and its place in and connection to the material world. What is essential for the truly ecological game, I argue, is acknowledging and actively *reducing* the harms involved in its own creation. This means becoming aware of the carbon emissions embedded in it through the game development process, the energy demands involved in getting it into players hands, the energy and emissions it entails from the players who play it, and what it facilitates or encourages in the form of the high-tech manufacturing sector's insatiable upgrade culture.

The ecological game shows us that a carbon neutral games industry must become an essential, non-negotiable part of its future. Without this, the best of efforts and intentions are shown to be a counter-productive waste of time, a syphoning of effort and energy if the industry does not also clean up its act. A carbon neutral games industry absolutely *must* be the critical, non-negotiable lens through which we imagine and plan the future of the games industry—anything less is a capitulation to our own demise, whether at the hands of stricter global regulation of emissions, changing public attitudes and acceptance, or worse. This should really be an uncontroversial perspective. We know that the world must transition rapidly from fossil fuels and other carbon intensive activity, and games are not exempt from this. They can no longer ignore their impact and the role they play in worsening the global climate crisis, however small that may (quite wrongly) be assumed to be. There are, however, numerous questions that follow from this, chief among them being how much of an impact does the global games industry have, in what capacity, and what can be done about it? The remaining chapters of the book set out to provide baseline estimates of just how carbon intensive the games industry is right now at the start of the third decade of the twenty-first century.

Chapters 4, 5, and 6 start this process by identifying emissions at key points along the game production process, from offices and workplaces, to the distribution of discs and digital files around the world, to game players' power consumption inside the home, and the very minerals themselves that are dug out of the ground, refined and transformed into gaming consoles and devices. At all stages we find emissions and other environmental harms that will need both rapid short term and deeper, lasting long-term action to address.

Chapter 4 reports on the findings of the very first survey of game developer's energy use and emissions, combining this data with the figures gathered from corporate sustainability reports from companies like Nintendo, Microsoft, and Sony. Together, this allows us to establish a very rough estimate from which to compare emissions intensity of game development at various scales, with kWh figures and carbon emissions for game developers ranging from single developers to employers of tens of thousands. This allows for the very first attempt at putting a rough figure on the total emissions of the entire games development industry around the world, responsible for as much as 15 million tonnes of $CO^2$ per annum, or

0.04% of global emissions. This figure suggests the game industry is more emission intensive than the entire global film industry, emitting about as much as the country of Slovenia. It is not all doom and gloom, however, and a number of encouraging case studies of game developers who are already reducing or offsetting emissions are explored, with examples of those currently aiming for, achieving, and even going beyond neutrality to carbon negativity.

Chapter 5 moves onto the question of games distribution—the emissions attributable to sending discs to the millions of players around the world—and the emissions that come from the players that play these games. Drawing on a small but growing body of literature that has conducted lifecycle assessments of different modes of distribution, from the earliest work by Mayers et al. (2015) that calculated emissions figures for PlayStation 3 games distributed to the UK in 2010, to Aslan's (2020) extremely detailed recent update which provides figures for the entirety of the European PlayStation 4 install base over its lifetime. Using what information is available about shipping and transport emissions, I undertake a case study of the emissions resulting from shipping discs to Australia, and estimate a single year's worth of emissions from disc-based game sales in the country. Comparing these with digital distribution's emissions has been a part of both previous studies, with each finding that current emissions from downloading games in certain cases exceeds those of disc distribution. Various factors affect this calculus at present, such as game file size, average download speeds, and the efficiency of internet communications infrastructure. I argue, however, that the *potential* for future decarbonization needs to be highlighted in any analysis and in planning for a carbon neutral games industry. From this perspective digital distribution becomes all but essential, doing away with significant transport emissions as well as plastic waste and other non-renewable materials. Looking at the incredible advances in renewable energy technology, pricing, and penetration and considering the potential (already realized in some parts of the world) for near completely emissions free electricity generation, it becomes clear that digital distribution of games is going to be a critical part of the carbon neutral games industry as these trends continue.

Chapter 6 surveys the existing literature on the emissions associated with play itself, collating energy use figures for different gaming devices, with the aim of providing game developers with the tools to estimate their own player's emissions. I draw here on industry leading work done by the first studio (to my knowledge) to attempt to completely offset its player's emissions, and the methodology they provide for how to do this relatively easily. To conclude the chapter, I consider what all this might mean practically for the practice of game design itself, how climate change and a zero emissions future might alter the way we design games. What do digital games look like after climate change? What new strategies could game developers, platform holders, and individual players embrace in order to curb their carbon emissions? I present some very initial suggestions in the hope that others take up this challenge and develop it further.

For all that there has been an increased focus on the material nature of digital games in the field of game studies (Apperley & Jayemane, 2012; O'Donnell, 2014; Whitson, 2018; Keogh, 2018; Nicoll & Keogh, 2019; Montford & Bogost, 2009; Gray, 2014; Freedman, 2018; Harper et al., 2018) there still remains a dearth of detailed and specific knowledge of the actual *material stuff* that makes up, or is used up in creating the devices on which we play games. Chapter 7 tackles this final, and extremely challenging part of the decarbonization of the industry, one which touches both the beginning and the end of gaming devices, and which plays a substantial part in determining the emissions intensity of games. While some understanding now exists among the general population that certain 'rare earth' and 'conflict' minerals are often present in the devices we play games on—thanks in large part to work like that done by Grossman (2006), Nest (2011), Merchant (2017), and Klinger (2018) as well as the reporting of many environmental journalists—specific knowledge about what exact elements are actually inside our devices, and what sorts of function they serve remains scarce. Much of the analysis of the problem of e-waste, for instance, has yet to meaningfully connect with the utility of the materials inside these devices, leaving them somewhat disconnected from the reasons *why* they are even used in the first place. Likewise, the focus on the *waste* dimension of e-waste can obscure the harms and emissions that have already occurred though mining, refining, and manufacturing. Chapter 7 presents the results of an inductively coupled plasma mass spectrometry (ICP-MS) analysis performed on the main system-on-a-chip of the PlayStation 4 console—the combined CPU and GPU that is referred to as the advanced processing unit or APU. The result of this test reveals the atomic components of the PlayStation 4's APU, and identifies the elements most likely to be present in large quantities. Chapter 7 looks at each of these elements in turn, considering how and where they are mined and refined, the carbon emissions and energy intensity of these processes, the harms for both human and other living organisms accompanying these extractions, and what role these materials might play in the component. I have titled the chapter the 'periodic table of torture' because it follows the hurt and pain generated at both the planetary and the personal levels via the extraction of these natural resources. It is also at least partly a speculative project, as is it almost impossible to know where, exactly, any particular element came from, and just as difficult to know precisely what the function of that element is. Best guesses and possible explanations are offered, and in the process, huge amounts of carbon dioxide emissions are encountered, alongside environmental contaminations, bioaccumulations of heavy metals, mountains of solid-waste burden, and a myriad other consequences for living things. All of them are connected to the PlayStation 4 gaming console, whether we like it or not. The choice of PlayStation 4 is as much for the device's popularity as our familiarity with it, seeking to destabilize and undermine the 'ordinariness' of this relatively cheap (and now, almost obsolete) high tech commodity. Millions of this device are now preparing to enter landfill in the coming years, but even before they get there they have contributed—however small a fraction—to the climate crisis. The chapter is ordered according to each element's number on the periodic table, covering sixteen specific metals found in substantial quantities within the tested portion of the device. Readers may choose to refer to this chapter as a reference section, perhaps when replicating this test with other newer devices, but it can also be read from start to finish, building up a rather bleak and troubling picture that shows just how difficult future decarbonization of hardware manufacturing may well be.

Finally, Chap. 8 concludes with a series of proposals for change, tailoring suggestions for different parts of the games industry, from game developers, to platform holders, to players themselves. Offering a series of actions for making beneficial changes to the games industry, ranging from the quick and easy (such as switching energy providers to 100% renewable or emissions offset electricity) to the much more challenging, like resisting and slowing down the hardware upgrade cycle through extending the lifespan of existing gaming devices. Some of the necessary changes will not be made lightly, so an essential part of this task is going to be the building of coalitions and organizing game developers to build collective power in their workplaces. It will almost certainly be a necessary step to completely decarbonize the games industry, as some of these changes may well present a challenge to profit margins. A democratically organized, developer driven climate movement, coordinating with civil society movements for climate action outside the games industry, stands the best chance of achieving the necessary changes. Alongside them will need to be game players and other figures in the industry—from journalists, to streamers, to researchers and activists on the industry's fringes—who together can make their demands be heard. It is becoming exceedingly clear all around the world that the powers of big capital and finance are not willing to make the requisite changes and investments in clean economies in the span of time we have. In order to forestall the worst of the climate catastrophe, we will need to *make them*. In order to do that, we will have to rediscover and recreate new levers of social and economic power, create new coalitions around shared interests among workers in and around the game industry in order to make our demands for a just transition, to decarbonize our workplaces and our entire lives.

The arguments outlined in this book are, like the climate crisis itself, complicated and at times sprawling. However they all unfold from the first and necessary act—taking the crisis seriously, of no longer holding separate our work and leisure from the biggest crisis we stand to face in our lifetimes. This is the kind of task that needs to happen across so many fields, so many disciplines, industries, and workplaces—but which I see increasing evidence of each day. Greater than ever numbers of people are reaching for achievable, grounded, and above all hopeful action—the kinds of change that start to move us in the right direction. Much like Morton (2010) noted, for some of us it is now shockingly, terrifyingly easy to see how we are all connected to the climate crisis, whether we like it or not. It is fast becoming the primary frame for much of our thinking and for many of our actions. The crisis we face is a massive one, seemingly endless in every direction, and at times it can feel like it is beyond the reach of any one of our efforts. Its urgency, however, is beyond question, and it will take all of our best, boldest, and most creative work to even begin to rectify. It will take coordinated political action to press politicians and entire economies to transition to fully renewable energy and thoroughly sustainable practices. But it is important for us to start now here, wherever we find ourselves, and with whatever we can currently reach for. It is my hope that these first steps outlined in this book expands our capacity, extends our knowledge, and allows us to achieve just a little bit more than we could before.

## REFERENCES

Apperley, T., & Jayemane, D. (2012). Game studies' material turn. *Westminster Papers in Communication and Culture, 9*(1), 5–25.

Aslan, J. (2020). *Climate change implications of gaming products and services.* Ph.D. Dissertation, University of Surrey.

Atwood, M. (2015, July 27). It's not climate change, it's everything change. *Medium.* https://medium.com/matter/it-s-not-climate-change-it-s-everything-change-8fd9aa671804. Accessed 30 Mar 2021.

Bastin, J.-F., Finegold, Y., Garcia, C., Mollicone, D., Rezende, M., Routh, D., Zohner, C. M., & Crowther, T. W. (2019). The global tree restoration potential. *Science, 365*(6448), 76–79.

Bennett, J. (2010). *Vibrant matter: A political ecology of things.* Duke University Press.

Bryant, G. (2019). *Carbon markets in a climate-changing capitalism.* Cambridge University Press.

Callon, M. (1986). Some elements of a sociology of translation: Domestication of the scallops and the fishermen of St Brieuc Bay. In J. Law (Ed.), *Power, action and belief: A new sociology of knowledge?* (pp. 196–223). Routledge.

Chess, S. (2017). *Ready player two: Women gamers and designed identity.* University of Minnesota Press.

Collings, D. (2014). *Stolen future, broken present: The human significance of climate change.* Open Humanities Press.

Davis, M. (2002). *Late Victorian holocausts: El Niño famines and the making of the third world.* Verso Books.

Dull, R. A., Nevle, R. J., Woods, W. I., Bird, D. K., Avnery, S., & Denevan, W. M. (2010). The Columbian encounter and the little ice age: Abrupt land use change, fire, and greenhouse forcing. *Annals of the Association of American Geographers, 100*(4), 755–771.

Dyer-Witheford, N., & De Peuter, G. (2009). *Games of empire: Global capitalism and video games.* University of Minnesota Press.

Federici, S. (2004). *Caliban and the witch: Women, the body, and primitive accumulation.* Autonomedia.

Flannery, T. F. (2005). *The weather makers : The history and future impact of climate change.* Text Publishing.

Flannery, T. F. (2017). *Sunlight and seaweed: An argument for how to feed, power and clean up the world.* The Text Publishing Company.

Freedman, E. (2018). Engineering queerness in the game development pipeline. *Game Studies, 18*(3), 1604–7982.

Gammage, B. (2011). *The biggest estate on earth: How aborigines made Australia.* Allen & Unwin.

Golding, D., & Van Deventer, L. (2016). *Game changers.* Simon and Schuster.

Gray, K. L. (2014). *Race, gender, and deviance in Xbox live: Theoretical perspectives from the virtual margins*. Routledge.

Grosz, E. A. (1994). *Volatile bodies: Toward a corporeal feminism*. Indiana University Press.

Grossman, E. (2006). *High Tech Trash: Digital Devices, Hidden Toxics, and Human Health*. Washington: Island Press/Shearwater Books.

Haraway, D. (1991). *Simians, cyborgs, and women: The reinvention of nature*. Routledge.

Harper, T., Adams, M. B., & Taylor, N. (Eds.). (2018). *Queerness in play*. Springer.

IPCC. (2018). *Global warming of 1.5°C. an IPCC special report on the impacts of global warming of 1.5°C above pre-industrial levels and related global greenhouse gas emission pathways, in the context of strengthening the global response to the threat of climate change, sustainable development, and efforts to eradicate poverty*. In V. Masson-Delmotte, P. Zhai, H.-O. Pörtner, D. Roberts, J. Skea, P. R. Shukla, A. Pirani, W. Moufouma-Okia, C. Péan, R. Pidcock, S. Connors, J. B. R. Matthews, Y. Chen, X. Zhou, M. I. Gomis, E. Lonnoy, T. Maycock, M. Tignor, & T. Waterfield (Eds.), In Press.

Joshi, K. (2020). *Windfall: Unlocking a fossil-free future*. NewSouth.

Keogh, B. (2018). *A play of bodies: How we perceive videogames*. MIT Press.

Klein, N. (2015). *This changes everything: Capitalism vs. the climate*. Simon and Schuster.

Klinger, J. M. (2018). *Rare earth frontiers: From terrestrial subsoils to lunar landscapes*. Cornell University Press.

Latour, B. (1993). *The pasteurization of France*. Harvard University Press.

Latour, B. (2005). *Reassembling the social: An introduction to actor-network-theory*. Oxford University Press.

Lorde, A. "The Master's Tools Will Never Dismantle the Master's House." 1984. *Sister Outsider: Essays and Speeches*. Ed. Berkeley, CA: Crossing Press. 110–114. 2007. Print.

Mayers, K., Koomey, J., Hall, R., Bauer, M., France, C., & Webb, A. (2015). The carbon footprint of games distribution. *Journal of Industrial Ecology, 19*(3), 402–415.

Merchant, B. (2017). *The one device: The secret history of the iPhone*. Hachette UK.

Methmann, C., & Oels, A. (2015). From 'fearing' to 'empowering' climate refugees: Governing climate-induced migration in the name of resilience. *Security Dialogue, 46*(1), 51–68.

Montfort, N., & Bogost, I. (2009). *Racing the beam: The Atari video computer system*. MIT Press.

Moore, J. W. (2015). *Capitalism in the web of life: Ecology and the accumulation of capital*. Verso Books.

Morton, T. (2010). *The ecological thought*. Harvard University Press.

Nest, M. (2011). *Coltan*. Polity.

Nicoll, B., & Keogh, B. (2019). *The Unity game engine and the circuits of cultural software*. Palgrave Pivot.

Nuccittelli, D., & Masters, J. (2020, December 21). The top 10 weather and climate events of a record setting year. *Yale Climate Connections*. https://yaleclimateconnections.org/2020/12/the-top-10-weather-and-climate-events-of-a-record-setting-year/

O'Donnell, C. (2014). *Developer's dilemma: The secret world of videogame creators*. MIT Press.

Parker, G. (2017). *Global crisis: War, climate change and catastrophe in the seventeenth century*. Yale University Press.

Pascoe, B. (2018). *Dark emu: Aboriginal Australia and the birth of agriculture*. Magabala Books.

Plumwood, V. (1993). The politics of reason: Towards a feminist logic. *Australasian Journal of Philosophy, 71*(4), 436–462.

Plumwood, V. (2009). Nature in the active voice. In *Climate Change and Philosophy: Transformational Possibilities* (pp. 32–47).

Rockström, J., Steffen, W., Noone, K., Persson, Å., Chapin, F. S., Lambin, E. F., Lenton, T. M., et al. (2009). A safe operating space for humanity. *Nature, 461*(7263), 472–475.

Rose, Deborah Bird, and Australian Heritage Commission. Nourishing terrains: Australian aboriginal views of landscape and wilderness. 1996.

Ruberg, B. (2020). *The queer games avant-Garde: How LGBTQ game makers are reimagining the medium of video games*. Duke University Press.

Smith, L. T. (1999). *Decolonizing methodologies: Research and indigenous peoples*. Zed Books Ltd.

Spufford, F. (2010). *Red plenty*. Faber & Faber.

Taffel, S. (2015). Towards an ethical electronics? Ecologies of Congolese conflict minerals. *Westminster Papers in Communication and Culture, 10*(1), 1–16.

Thomas, C. D. (2017). *Inheritors of the earth: How nature is thriving in an age of extinction*. Hachette UK.

Wainwright, J., & Mann, G. (2018). *Climate leviathan: A political theory of our planetary future*. Verso Books.

Whitson, J. R. (2018). Voodoo software and boundary objects in game development: How developers collaborate and conflict with game engines and art tools. *New Media & Society, 20*(7), 2315–2332.

Yusoff, K. (2018). *A billion black Anthropocenes or none*. University of Minnesota Press.

Zalasiewicz, J., Waters, C. N., Summerhayes, C. P., Wolfe, A. P., Barnosky, A. D., Cearreta, A., Crutzen, P., et al. (2017). The working group on the Anthropocene: Summary of evidence and interim recommendations. *Anthropocene, 19*, 55–60.

# CHAPTER 2

# How Can Games Save the World?

I suspect the first question many in and around the games industry have, the first place they start from when thinking about climate change is this: what can games do to help? It's a natural question and a good place to begin. The desire to use our own skills and expertise to contribute to the biggest challenge of our time is a noble one. Initial responses, particularly for game designers, will probably be further questions—perhaps around what sorts of games can be made, or what can be done through games to persuade or convince players of the urgency of climate action. Can we make games that convince players to take the climate issue more seriously? Can we use games for better urban planning to avoid heat island effects? Can we embed green themes and ideas into games? Can we design games to get players engaged in good habits, to act on sustainability, or take steps to lower their carbon emissions?

There are two problems with this approach. The first being that it begins (understandably) from what we are already familiar with—the games themselves—causing us to overlook other, simpler or more effective questions and their answers. The most immediate cause of climate change is $CO^2$ in the atmosphere, and there is already plenty that games can do about that. Putting that aside for now (we will return to it in Chap. 3 onwards) there remains another issue, namely that these sorts of questions assume a singular, homogenous and uncomplicated sort of player. In this chapter I examine the literature on games and their effects,

B. J. Abraham, *Digital Games After Climate Change*, Palgrave Studies in Media and Environmental Communication, https://doi.org/10.1007/978-3-030-91705-0_2

particularly regarding the much-discussed potential for games to persuade, advocate for an issue, or create social change, arguing that there are several good reasons to temper our enthusiasm and pause rush to deploy games in this way.

Perhaps the game industry's best-known evangelist for the power of games to change the world is Jane McGonigal. Her 2011 book *Reality is Broken* quite literally argues that games are powerful (and appropriate) tools for reconfiguring everything that ails us, even the world itself. She opens with the following series of rhetorical questions:

> What if we decided to use everything we know about game design to fix what's wrong with reality? What if we started to live our real lives like gamers, lead our real businesses and communities like game designers, and think about solving real-world problems like computer and video game theorists? (McGonigal, 2011: 7)

The trouble with McGonigal's vision in *Reality is Broken*, and which itself is emblematic of a wider tendency that we shall encounter in this chapter, is that it is largely a self-disciplinary, or self-help project, and like many enthusiastic accounts about games various and unique powers sees in every problem games as their potential solution. For the most part, the targets of McGonigal's interventions are individuals and their perceptions, with little if anything to say about how to deal with a world filled with recalcitrant stuff and powerful global actors—things that do not much care about imposed rules or human perceptions. This becomes clear when looking at the games held up by McGonigal (2011) as exemplars. In analysing the game *Chore Wars* for example—which gives players XP for completing household chores and invites playful min/max-ing of drudgery like cleaning toilets—she claims, based on her household's adoption of the game, that it has "has changed our reality of having to do housework, and for the better" (McGonigal, 2011: 121). But what has *Chore Wars* actually changed? The chores remain, just with a game-like scaffold around them. All that has changed is, at best, how the players feel about doing them, perhaps how motivated they are to complete them. This sort of brain hack approach to solutions is immediately inadequate when applied to anything more complex than simple chores. Similarly, McGonigal's *SuperBetter* game adds extrinsic game incentives and structures, wrapping a super hero theme around tasks already well established by the medical literature to speed up recovery from certain illnesses (McGonigal's test case was her own protracted recovery from an unexpected concussion). As a solution for a problem of human motivation, perhaps even a problem of creating health-promoting habits and routines *for an individual* this is fine enough. For problems where a lack of motivation is the main barrier, adding points or an incentive structure may well be an appropriate, even excellent, solution. But for more tenacious, complicated problems with non-obvious, external or material causes involving a level of complexity beyond the scope of an undisciplined individual human, it is difficult to see how adding such an arrangement could possibly work. Adding gold stars next to the names of people who reduce their carbon emissions is not a solution, it's a distraction at best, coming at the time when we most need to be focussed on actual results and concrete emissions reductions.

Even for situations where the problems *are* entirely located within a single individual, it does something of a disservice to these people to think that every one of them could be (or should be) motivated by simple (or even complex) game-like structures. As we shall see, this flattening of "the player"—typically done through an enthusiastic framing that is keener to account for the design of games than their real live players—is a common, though one-sided feature of much of the writing about games and the uses they can be put to. When we actually stop to consider the breadth of views, deeply held beliefs, embedded ideologies and individual opinions held by players, we begin to see how difficult these can be to interrogate, and the paucity of much of the work around games purported power to make change. The same insight guides us toward the realisation that issues that might, on the surface, seem to stem from individuals lacking motivation are inevitably more complicated, being historically and materially instantiated in the wide context of a human life. What might work for a small cohort of Silicon Valley academic consultants, might fall laughably flat when transplanted to a very different social context, a very different lifeworld. The Rohingya in Myanmar are unlikely to get much out of an online game incentivising the ploughing of fields or the planting of crops. Likewise, the South Sudanese concussion-patient who lacks access to appropriate health care, owing to remoteness, poverty, or war is unlikely to be able to reap the proposed benefits of getting *SuperBetter*. All of this complexity is steamrollered by the enthusiasm for games unmoored from the messy, complicated reality of living, breathing, thinking and feeling players.

Even if it were possible to flatten players into a mostly-homogenous mass of willing climate gamers, climate change for the most part does not require more positive perceptions or even a more 'motivated' group of individuals to solve. In a diagnostic sense, climate change requires quite simple solutions: having less $CO^2$ and other greenhouse gases in the atmosphere. But being able to state that, however, gets us no closer to achieving action and change capable of untangling the myriad systems of entrenched power, big finance, histories of racist colonialism, captured political systems, and fossil fuel cultures that have become embedded over decades, even centuries (Moore, 2015; Malm, 2016; Yusoff, 2018; Huber, 2013). Are games *really* powerful enough to cut through such a tangle, one that has eluded the world's best scientists, its smartest and most motivated organisers, its most staunch activists, and some of the biggest most globally coordinated social movements in recent history? We should have some humility in the face of the task before us instead of immediately jumping to our favoured games as a solution.

This chapter begins by looking at the body of existing literature on climate games and the proposed utility they have to help us face this challenge. I characterise this literature as overly optimistic, suffering from a one-sided focus on games and game mechanics and their "potential"—never far off, but rarely, if ever realised—at the expense of an understanding of the players to be reached. According to this literature, games can be used for climate education, to help improve public governance and planning, to help us inhabit future scenarios, and to teach sustainability principles. Games may be able to do some, or even all of these things, but under what conditions, and for what groups or individuals are these effective? By considering the existing literature on games potential to persuade, to change the minds of players in one way or another, we find important reasons to be cautious in affirming the suitability of games for many of their proposed ends. Thinkers like James Paul Gee (2003) and Katie Salen Tekinbaş (2008) present important reminders to pay attention to the agency of players—even resistive or recalcitrant agency—and to not take them or the lived contexts we find them in for granted.

One of the effects of describing what games 'can' or 'might' be able to achieve without reference to a particular player, is to gloss over or downplay much of the barriers to be overcome in achieving, for example, climate change persuasion. The chapter thus discusses in particular Ian Bogost's (2007) concept of simulation fever—part of his broader argument about the procedural rhetoric games are able to achieve—which holds important space for player resistance and disagreement, remaining never quite possible to entirely close off players ability to reject the premise or results of a mechanical 'simulation' (of climate change, for instance). I see this space being amplified by the resisting effects of climate scepticism and related ideology, a problem that becomes especially pronounced for climate games. In the rest of the chapter I argue for the critical importance of considering the challenges presented by ideology specifically—and the need to reassert the centrality of the mystified player and their internal process of meaning making in our analyses. This perspective is informed by best practices in the field of climate communication, which emphasises the complexity of the task, underscoring the importance of an 'interaction' model in climate communication.

The question for digital game makers who wish to aid global responses to the climate crisis becomes rather different, becoming more important than ever to ask how might games perform ideological critique. Is it even possible to get around ideological barriers and avoid the conscious rejection of climate games? I suspect that it may be, based on a reading of a rather unexpected source—the military simulation game *ARMA 3*. However, rather than engaging with the climate through modelling its mechanical systems, via simulation, or at the level of factual debate, I find *ARMA 3*'s creation of an aesthetic vision, offers players a sense of what a renewable-energy powered future might feel like, placing the player inside different material-ideological coordinates. The same caveats still apply, however, and I have no evidence that the game has ever convinced anyone. However I argue that, at the very least, the game's use of an aesthetic vision of climate change avoids both the paralysing effects of 'the sublime' which often accompanies climate change art, as well as the conscious rejection of particular simulations of our climactic future.

## GREEN AND CLIMATE GAMES

Over the last decade a small but growing body of work has appeared focussing on the proposed uses of games in confronting and combatting various aspects of climate change. Much of this work looks to identify, describe or analyse the positive impacts that games can make as agents of change, or their ability to promote wider environmental issues and consideration. Here I survey some of these claims and what they suggest games potential to be.

Much of the work in this area is concerned with reproducing or simulating climate or environmental dynamics in games. One of the earliest examples of this, and emblematic of this sort of approach, is an off-hand comment that appears in Barton's (2008) discussion of the history of games simulation of weather effects. Barton (2008) argues for greater environmental verisimilitude and the deliberate inclusion of environmental simulations in contemporary games, asking rhetorically: 'how can games acknowledge the threat of global warming when game characters fail to take notice of a torrential downpour on their heads?' What if we don't treat this as a rhetorical question, and consider directly this assumption. Does game characters ignoring weather prohibit wider acknowledgement of the climate? As we have seen already, climate change is not interfaced with by humans directly, through rain falling on our heads. It is precisely the scale on which climate change operates which places it beyond our direct experience—both in its temporal duration, its spatial distribution, and its political, economic, and material-stratospheric causes. Even those of us who have lived through the increasingly common disasters made worse by climate change are not directly experiencing it. Climate change itself resists and evades our experience—it is more akin to increasingly weighted dice than a finger that squeezes the trigger of a gun. Even climate science only ever encounters it mediated through experiments, measurements and supercomputer modelling—how much more-so for those of us who get it through news reports and increasing horrifying anecdotal experience. The assumptions that Barton's rhetorical question rests on are that (a) direct experience of global warming is possible in general, and (b) it is possible through game mechanics in particular. But what possible game *mechanic* or set of mechanics could ever hope to reproduce the entirety of such a vast problem?

The other feature of much of this work is a frequent, and I would suggest perhaps unfounded, optimism. An early example of this is the framing of findings from the Pardee report out of Boston University from 2012, which discusses the potential for games to assist players understanding of climate modelling and decision making, claiming that 'participatory games can help us "inhabit" the complexity of climate risk management decisions, allowing us through system dynamics modelling to explore, then test a range of plausible futures' (de Suarez et al., 2012: 6). The question of what sort of player would be interested in or would gain something from "inhabiting" these complex dynamics is left unasked, however, an issue that turns up in other reviews of the literature (Galeote et al., 2021). Perhaps elected world leaders might benefit from playing these sorts of games, but are they willing or able to do so? And does this sort of thing seem appealing to the type of gamer who needs to be convinced to take notice of the environmental impacts of their own gaming? Who are these games for?

There is no shortage of work in this vein that couches its argument in terms of possibilities or potential—that games "might" or "could" do something. The Pardee report's outlook is similar to McGonigal's enthusiasm for the 2007 participatory game *Word Without Oil* which, she writes, 'was the proof-of-concept game that convinced me we really can save the real world with the right kind of game' (McGonigal, 2011: 312). That game involved about 1900 people recruited largely from her online networks of interested self-selected volunteers, according to McGonigal (2011: 306). Though her discussion is filled with glowing testimonials, there is little evidence that the project could ever be successfully scaled up, or how it might be made to work for non-volunteers. A small, group of interested believers poses quite a different challenge than a group marked by apathy, the former might reasonably be described as preaching to the choir, hardly cause for celebration in the struggle for real climate action.

Likewise, Bell-Gawne et al. (2013) have argued for the potential of games' simulations to change or influence player beliefs about environmental policy. They argue that games 'allow us to preconceive the results of [social or environmental] policies over consecutive trials and foresee the quantitative effectiveness of such policies' (Bell-Gawne et al., 2013: 94). One does have to ask, however, how accurate do these simulations have to be to provide such insights? How do these 'simulations' differ from or improve on existing public policy research methods? Does each game need tailoring to the specific needs of the particular public, the particular locality and situation to be simulated? How could this be feasible at scale? Such questions are rarely raised. Once again, the potential utility signalled here becomes somewhat less exciting when considering what sorts of players stand to benefit from, or even be interested in engaging with, these sorts of experiences. Does a gaming-ready audience of public policy planners even exist?

A 2013 survey of multiple forms of digital, board and card games that address climate change conducted by Reckien and Eisenack (2013: 266) found 52 games involving climate themes or climate change elements. They argue, based on their analysis of the games and the number identified, that games about climate change 'are not a niche product anymore' and 'make a valuable contribution to efforts to look for solutions to [climate change]' (Reckien & Eisenack, 2013: 266). Slightly more recently, Wu and Lee (2015: 414) have performed a similar review of climate change games, finding that 'a significant number of online climate change games exist as mini-games or simple simulations' and that 'these are generally found on websites geared towards younger audiences.' Despite this, the authors maintain some level of enthusiasm, and paint a fairly rosy picture for climate games, claiming they 'offer powerful tools for education and engagement' (Wu & Lee, 2015: 413).

Another example in the 'games as simulations' vein is the argument offered by Kelly and Nardi (2014) for the potential of games to lead players through 'imaginative visions of situational potentials and solutions to problems.' Even here, however, the arguments tend towards the design or mechanics-centric approaches to explaining games' powers in an attempt to reach a relatively strong and uncomplicated position on games' efficacy. Kelly and Nardi (2014) argue that games can have players engage with problems like resource scarcity and the necessary social changes these situations suggest which, while probably true, follows the same structure of simplifying and flattening the role of the player. But considering what games *can* do often happens without reference to what they what *actually* do, extrapolating a universal power to act upon a largely faceless, largely homogenous, and always desubjectified "player". By focussing on the game while downplaying the reality of the player, it becomes plausible to link specific arguments about games of a certain type with their supposed outcomes upon a homogenised player population. I should stress that this is not something I am singling out Kelly and Nardi (2014) for in particular, as it has become a trope of much game scholarship to employ this approach. By focussing on specific mechanics divorced from real game playing subjects it's possible to make claims like the following about games provoking reflection or conscientious engagement with resource scarcity scenarios: 'good game mechanics can cultivate imaginative visions of situational potentials and solutions to problems' (Kelly & Nardi, 2014). Here, as in all the above literature, there's a lot of work being done by the word 'can'. As is often the case with this sort of argument for one or another of games persuasive or interventionist powers, by not identifying who the players are we might expect to experience this effect, many problems are smoothed over. Thorny issues of unreliable reception, of user experience and design clarity, of clear communication of intentions, let alone getting the right games to the right people, all are kept out of the picture. It places potential detractors in a difficult position as well—wanting some certainty about what games *do* rather than *can* places us in the awkward position of appearing unenthusiastic or sceptical about games, a wowser, a downer, or simply a *critic*. Wanting some certainty in our climate tactics efficacy, however, should be mandatory at this point. Good intentions are not going to cut it.

Kelly and Nardi (2014) go on to argue that sustainable practices 'could provide material for the thematic and aesthetic design elements of new games. Global futures games can make visible the possibility of low/no growth as a challenging and achievable goal' (Kelly & Nardi, 2014). While these are laudable goals, and I too want see these issues taken up by the wider population, by leaving unstated who needs this sort of intervention, it becomes hard to disagree with without being charged as an uncharitable reader. Many climate (and games) researchers would surely agree to seeing value in the search for new, exciting and productive approaches which have encouraging results on players and their gameplay experience. But around games, for whatever reason, such exploratory work often takes on an expansive and enthusiastic character. It is an enthusiasm that becomes harder to justify when encountering the barriers, agency, intransigence, and ideological resistances of the very real, and not at all predictable or homogenous game player. The hypothetical player in much of this literature could described considered a 'designed identity', much how Shira Chess (2017) describes in *Ready Player Two* the construction of a certain sort of 'female gamer' identity, counterpart to the assumed but unstated male "gamer" identity. The designed identity of the female gamer is 'a hybrid outcome of industry conventions, textual constructs, and audience placements in the design and structure of video games' (Chess, 2017: 5). The designed identity of the invisible, undifferentiated "player" in much of the literature on games potential for change or persuasion is similarly constructed—the illusion of a real person whose complicity with the game's particular ends can be safely assumed.

In a recent meta-review of 14 different studies into the uses of gamification techniques in a climate communication context, Rajanen and Rajanen (2019) argued once again for optimism about the use of gamification in teaching climate or environmental skills, developing climate knowledge, and so on. Concluding that, as a result 'climate change gamification is a research area that deserves more attention' (Rajanen & Rajanen, 2019: 253) they at least acknowledge that 'longitudinal studies were missing from the employed approaches, while these are especially needed to assess to what extent the learned skills are applied in real life' (Rajanen & Rajanen, 2019: 262). One of the precious few longitudinal studies to test the effectiveness of gamification approaches was recently completed by

Raftopoulos (2020), showing significant drop-off in the applied use of gamification from early adopters. It should perhaps not surprise us that in real world application, of an initial 23 organisations that were making use of gamification techniques, between 2014 and 2018 follow-up contact found that '60% of organisations reported that they no longer use gamification in their work, 22% said they were using it less, 9% stated it was about the same and only 9% said they were using more of it' (Raftopoulos, 2020: 3–4). When the heterogenous reality of actual players meets the game, clearly things are not as simple as some of the literature would have us believe. How much more so will this be when it comes to climate action—a topic that is often hotly contested?

Not all the literature on games and climate and environmental topics suffers from these same issues, however. Recent work bringing ecocritical perspectives to the study of games touch on similar concerns without necessarily replicating all these same issues. Player engagement with a complex environment is taken up by Kyle Bohunicky (2014) who argues that in *Minecraft* (2011–), players experience and act out greater environmental agency than is typical. This includes engaging in the writing of terms like "shelter" or "transportation" onto the environment, arranging different block materials into buildings, paths, and other structures. Again, we still need to ask who these players are or could be, though at least here the unspoken player operates as a stand in that allows for description rather than expansive categorical claims. The primary claim Bohunicky (2014) makes is that the survival–crafting genre has a unique relationship with nature and ecologies, an issue that I discuss in Chap. 3 specifically, evaluating the survival–crafting genre and its claims to greater ecological complexity and fidelity. Bohunicky (2014: 231) however centres the player in this analysis, rather than the game object or game mechanic, noting the effects of player agency and their capacity to act upon the procedurally generated "natural" landscape—comparing this to the act of 'writing', translated to a digital game context.

Approaching some of these same problems with substantially more nuance and care for material realities is the pioneering work of Nick Dyer-Witheford and Greg de Peuter (2009). In their now canonical work *Games of Empire* they examine games' entanglement with the processes of global capitalism and empire building. Evaluating the genre of "policy simulator," citing at least seven games about faming, climate, weather, business, capitalism, and so on, they note these games' questionable environmental credentials:

Most code neoliberal assumptions: Food Force, for example, engages players with issues of global famine but never really probes the structure of the world market. Other serious games are sponsored by flagrantly hypocritical corporate philanthropy. (Dyer-Witheford & de Peuter, 2009: 201)

They also see some value in 'serious games' not produced by compromised actors but by activists, hackers and countercultural figures:

the compromised nature of many current serious games does not mean the genre lacks radical potential. Eroding the monopoly of the military-industrial complex over simulation tools, however modestly, to foster their use by ecologists, peacemakers, and urban planners, is a welcome development. (Dyer–Witheford and de Peuter, 2009: 201)

Dyer-Witheford & de Peuter (2009) also raise the issue of the carbon footprint of the games industry, in one of the first attempts that I am aware of to put an actual figure on the carbon intensity of a given game activity. Regarding the online game *Second Life*, they note:

your personalized avatar is powered not just by mouse clicks but by computer servers that, according to one estimate, annually use about 1,752 kilowatts of electricity per Second Life resident, as much as is consumed by an average actual Brazilian, and generating about as much $CO^2$ as does a 2,300-mile journey in an SUV. (Dyer-Witheford & de Peuter, 2009: xii)

This acknowledgment of carbon emissions represents a far more intimate connections between games and the climate crisis, and a thoroughly welcome turn. Later chapters of this book attempt to extend this impulse to as much of the industry as possible, but their initial figure represents a small but important interest in the materiality of games and their global impact, which has since been taken up by others. (Guins, 2014; Newman, 2012; Apperley & Jayemane, 2012).

Perhaps the most prominent work, and some of the most thorough scholarship in the area of games studies and environmental and climate issues has been written by Alenda Chang, who was writing as early as 2011 that 'games naively reproduce a whole range of instrumental [human-environment] relations that we must reimagine' (Chang, 2011: 60). This is an important insight, and a serious weakness within western thinking and western environmentalism more broadly, tending to reproduce what Latour (1993) describes as the hopelessly flawed nature/culture split, with the world or nature remaining 'out there' only where it remains pristine, untouched by human influence. Chang's (2019) book *Playing Nature* directs us to Morton's notion of a 'dark ecology' instead, which

> reminds us that ecology is not solely about the bright optimism of interconnection and interdependence, a warm, furry, mammalian comfort in our cohabitation, but also a universe of waste, dirt, shit, and trash that does not disappear, though it may fade or become otherwise as it gets taken up again and again by a sprawling web of organisms and inorganic actors. (Chang, 2019: 173)

Chang (2019: 179) wants to see a 'dark ludology' emerge in the form of games acknowledging the waste by-products of in-game activities and other ecological processes. By this point it should be clear that I am not particularly swayed by representational or mechanical content of games matching this or that environmental or ecological process, and so while acknowledging these waste by-products is an important first step, this cannot simply be done representationally. Chang (2019: 146) however is still interested in seeing games do better at 'demonstrating the flow of energy and material through human and nonhuman systems, which would in turn underscore such core ecological premises as limitation and unpredictability.' For the reasons I am outlining in this chapter, I do not think this needs to be high on the agenda for a climate conscious game industry. For me, this sort of work tends to distract, and all too often syphons away effort and attention that could be better spent working on actual carbon emissions reduction. As climate games activist and organiser Hugo Billie often notes, however, climate games are almost certainly coming anyway. Games do tend to reflect the concerns and interests of their makers, and we are entering an era where the climate is becoming the greatest concern for many of us. In that respect it may be worth knowing what to avoid, for instance, when making climate games.

But that work cannot come at the cost of distracting us from the most pressing concern which is the necessity of real material improvements and concrete action on emissions and other environmental harms that are needed in the world of 'waste, dirt, shit and trash' that Chang (2019) describes. Her work on this under-appreciated aspect of games is illustrative, and shows the broader shift in thinking that is sorely needed. Chang (2019: 153) writes that her work seeks 'to reaffirm that games and gameplay occur in our world and therefore when we play, heat and energy are exchanged as much as data and social communication.' The focus on heat and energy, waste and data, as well other 'neglected material aspects of games and game platforms and... their inevitable demise' (Chang, 2019: 146) is an important, and far more pragmatic way that games can be part of saving the world. In comparison to much of the work that wishes to use games to save the world through changing hearts and minds, cleaning up the waste and other material impacts of games is going to be so much easier.

## How Do Games Change Minds?

There is no shortage of theoretical approaches seeking to explain games' persuasive, activist, or even pedagogical potential, with many drawing upon a number of different frameworks for understanding this question. As we have just seen, however, many rely on or reproduce the same impulses as McGonigal (2011), replicating issues that foundational media studies research has either cast doubts upon or disproven. In this section I briefly canvas the literature on games educational, pedagogical, and persuasive potential in more detail, finding further evidence for the necessity of a complicated picture of 'the player' in our analyses.

At the start of the 2000s, after media and politicians in the United States began turning their attention to videogames as scapegoats for teen violence and school shootings, game players, developers, and even many scholars began resorting to defensive arguments in order to dissipate the negative attention building around the games and violence debate. While understandable in context, and having now been largely upheld by two decades worth of research that has found no lasting impact of game play on levels of violence (Elston & Ferguson, 2014; Szycik et al., 2017) it nevertheless placed many on the back-foot, and lead to thorny questions for those who wanted to talk about the *positive* effects that games can have further down the road. The accidental effect of rhetorically closing the door to violence, at least as a simplistic "effect" of games, was that it became hard to hold open the possibility of games many *positive* consequents. Because if they don't *cause* violence, what *do* they do to players?

Perhaps the best work from this period and its aftermath remains those from scholars who saw this bind and avoided knee-jerk responses to the videogame violence debate, steering clear of simplistic reduction to debunked 'media effects' models like the hypodermic needle theory (wherein media injects the message into the receiver) entirely in favour of more complex approaches. The work of James Paul Gee (2003) is still some of the best, most foundational work on games potential to impact players, and particularly to *teach* players, recognising the importance of a range of contextual factors that come into play. Focusing primarily on players and the contexts in which they engage in "learning", Gee (2003) employs a definition of learning that is active, critical, and engaged (a far cry from the passive consumption of facts or the accumulation of information) resulting in a figure of the player as an active agent in their own learning. The conclusion leads Gee (2003: 46) to affirm that while 'video games have the potential to lead to active and critical learning', players will always have some say in this. He asks, crucially, 'what ensures that a person plays video games in a way that involves active and critical learning and thinking? Nothing, of course, can ensure such a thing' (Gee, 2003: 46). We should take this observation to heart, with awareness of this often elided in the more enthusiastic theories and accounts of how games influence, effect, change or persuade players and society writ large, as we have just seen. Gee's (2003) perspective underscores the necessity of an individual's active engagement and participation in the process—their willingness to engage with and interrogate ideas, assumptions and their own mental models—and points towards a little discussed gap between intentions and outcomes. This gap looms even larger when considering some of the possible internal resistances and barriers that reside within the to-be-convinced player. Not least of these resistances involves ideology itself, perhaps the quintessential challenge facing any game that seeks to persuade on the issues around climate change.

Along similar lines, in her introduction to *The Ecology of Games*, an edited collection focussing on games' ability to facilitate change or learning in children and younger players, Katie Salen Tekinbaş offers the following summary of the existing discourse in this area, which 'has been, to date, overly polemic and surprisingly shallow' (Salen Tekinbaş, 2008: 2). The "value" of games and their ability to challenge cultural and ideological constructions has been mired by extremes of both panegyric defensiveness and the (itself ideological) "neutral tools" discourse freighted in from certain fields of science and technology, despite having been widely complicated and criticised therein (Haraway, 1991; Latour & Woolgar, 1986; Winner, 1986; MacKenzie & Wajcman, 1985).

In Stevens et al.'s (2008) chapter in the same volume, the authors contribute an important framework for understanding the assumptions of much research into the effects of games on players, rejecting what they describe as a "separate worlds view" of games and their impact on players.

They offer a research method influenced by 'situated, everyday, or distributed cognition' (Stevens et al., 2008: 42) that looks at ordinary everyday situations in order to have a 'basis to credibly claim that our research accounts are about how and what people do, learn, and think in daily life, and not simply about what they do within the context of contrived laboratory tasks.' By interrogating this "separate worlds view," and drawing upon the concept of "transference," itself a contentious term describing the application in one domain skill or knowledge gained from another (e.g., skills learned in games applied in "the real-world"), Stevens et al.'s (2008) are able to account for the young people's game playing diversity. They note that 'the culture of game play is one that is quite tangled up with other cultural practices, which include relations with siblings and parents, patterns of learning at home and school, as well as imagined futures for oneself' (Stevens et al., 2008: 43). Their conclusions preclude simple answers about the "effects" of games on their players, particularly in the now ubiquitous cultural context of constant adaption, remixing/repurposing and the like. For those asking what effect games have on the young people that play them they offer only the following:

> an "answer" to the question of how media consuming and repurposing has affected these young people is complicated and contingent; it depends on differing dispositions and purposes that people bring to play, who they play with, and perhaps more importantly what people make of these experiences in other times and places in their lives. (Stevens et al., 2008: 63)

This emphasis on externally contingent dimensions of individual player's lives is often missing from many accounts that focus on games and their "potential". Stevens et al. (2008: 83) note with their conclusions that they are deliberately "stepping quite far away from any simple generalizations about effects of video game play." A similar emphasis on the contingency and specific materiality of any given moment or context of play is found in Apperley's (2011) concept of 'situated gaming'. For Apperley, such an approach rests on two key principles. The first is that 'the materiality of the embodied experience of gaming' is in some way affected by 'local cultures and contexts of play' (Apperley, 2011: 35). In other words, in different places, different cultures and contexts bear down on the game experience, not necessarily determining it, but shaping it and any processes of meaning-making entangled with it. The second principle of situated gaming (and all gaming is 'situated' gaming) is:

that the game experience is played out as a negotiation between the 'global' immateriality of the virtual worlds of the digital game ecology and the myriad material situated ecologies that are manifestations of the 'local'. (Apperley, 2011: 35)

Stevens et al.'s (2008) ambivalence over generalizable conclusions around game effects in children and young people is mirrored in Apperley's (2011) acknowledgment of the substantial differences that can and do exist around the world in cultures, habits, and material circumstances—in other words, differences in the many different contexts in which play happens. This is not a problem unique to youth, more pronounced though these may be for persons just starting out in life. We ought, therefore, to extend some of the same concern to our accounts of the adult population—yet there is little such modesty to be found in much of the related literature on games and the "learning" opportunities they present to players, even those players that have developed and established far more ingrained intellectual and ideological habits and proclivities. As in the climate games literature discussed above, the efficiency with which games' persuasive powers are proffered seem incompatible with theories of players own agency and the necessity of their active participation in meaning making deriving from game experiences. In the following section, I turn to examine in more detail the role of ideology, the barriers it presents to climate action in games especially, and the lessons to be learned from climate communication research.

## IDEOLOGY IN GAMES

In his landmark text Persuasive Games, Ian Bogost (2007) addresses the question of games ability to perform ideological critique and other kinds of persuasion and illumination. Bogost begins with a brief history of theories of ideology, noting that, 'hidden procedural systems that drive social, political, or cultural behaviour are often called ideology' (Bogost, 2007: 72), and this formulation of ideology crucially connects to his overall argument about games efficacy in enabling players to grapple meaningfully, even critically, with systems or procedures. Bogost traces a short history of the term "ideology" across a number of not always entirely reconcilable thinkers and traditions. Beginning with its etymological origins with Antoine Destutt de Tracy as a 'science of the origin of ideas' (Bogost, 2007: 73), he also canvasses Marx's conclusions about ideology

('ideology entails the delusion that ideas are material') to Althusser's modification of Gramscian conceptions of the term. Bogost finally settles, perhaps reluctantly, on Zizek's approach to ideology, largely a materialist one. In Bogost's estimation, 'Althusser essentially collapses the realm of ideas completely into material practice' (2007: 74) with his focus on ideology's instantiation in apparatuses, while for Zizek:

> Ideology remains material… but this material reality is distorted and malignant. Ideology is not just a false representation of reality, it has become a part of reality itself, disfiguring it. (Bogost, 2007: 74)

As a consequence, Bogost concludes that for games, as in other forms of political and rhetorical persuasion, 'the challenge that faces political critique, then, is to identify the distortion in material practice' (Bogost, 2007: 74). I find this a somewhat limiting approach which treats ideology more like an engineering problem to be overcome than something involving unpredictable and irreducible human complexity. Critiquing and unpacking ideology often touches on deeply held, intensely personal beliefs, and can take the form of a deep abiding struggle over questions of great personal significance. Nevertheless, Bogost's analysis retains a valuable insight, through its emphasis on the space for individual resistances and recalcitrance, as might occur when a player discounts ideas ideologically opposed to their own. Bogost reaches this through his concept of "simulation fever":

> The disparity between the simulation and the player's understanding of the source system it models creates a crisis in the player; I named this crisis simulation fever, a madness through which an interrogation of the rules that drive both systems begins. (Bogost, 2007: 332–333)

Although Bogost only envisages simulation fever occurring when a player resists the procedural instantiation of a real-world system unfaithfully reproduced, leading (though perhaps not necessarily) to "interrogation", I want to suggest that simulation fever may be usefully re-cast as resistance to a particular procedural implementation for any reason. Simulation fever does not need to be constrained to an internal response to an inaccurate picture of the world, instead it might encompass any resistance to the way a procedure operates. Simulation fever for Bogost appears to be a potentially productive start to the interrogation of the players own mental models, however I think it is necessary to remember Gee's (2003) caution that nothing can ensure a player plays in a critical, active, and reflective way—so much more the ideologically mystified player.

Further, how would we get around or avoid the problem of subjectivity when it comes to the accuracy and veracity of our own evaluations, our estimations of the faithfulness of a procedure to the so-called "real-world"? As the work of Max Horkheimer (1947) in particular, and the Frankfurt school of critical theory emphasises, the historically constructed (and not at all 'natural') habits and regimes that form our very own sense perceptions themselves thoroughly complicates any potential claims to viewing the world "as it simply is". How could we evaluate verisimilitude or faithfulness of reproduction without invoking our own ideological and historically contingent predispositions—or without invoking a naive realism, just as ideological a vision of the world itself?

From Bogost's brief history of ideology, however, we can pick out elements of a more developed analysis of ideology. Terry Eagleton (1991) describes in detail two distinct, but not entirely separate, traditions of thought regarding ideology:

> One central lineage, from Hegel and Marx to Georg Lukacs and some later Marxist thinkers, has been much preoccupied with ideas of true and false cognition, with ideology as illusion, distortion and mystification; whereas an alternative tradition of thought has been less epistemological than sociological, concerned more with the function of ideas within social life than with their reality or unreality. (Eagleton, 1991: 3)

Eagleton (1991) suggests that most common uses of the term "ideology" (like Bogost's reading of Zizek) more resemble the former "mystified" perspective than the latter, noting that to claim something or someone as 'speaking ideologically is surely to hold that they are judging a particular issue through some rigid framework of preconceived ideas which distorts their understanding. I view things as they really are; you squint at them through a tunnel vision imposed by some extraneous system of doctrine' (Eagleton, 1991: 3). Yet it remains inconceivable that anyone could successfully claim to say they see things free from *all* distortion, and in this way, all seeing becomes ideological.

There are no simple answers to this problem, and it is one philosophers have argued over for centuries in various guises—from Plato's shifting patterns on cave walls to more contemporary thinkers, like Quentin Meillassoux (2008) who argues for a belief in *the absolute* via mathematical theory and a rigorous argument for the utter necessity of contingency itself. For our purposes however, the problem posed by ideology, truth and perception poses is perhaps best elaborated by Horkheimer (1972) himself, who approaches the question of knowledge and truth by emphasising the aforementioned *construction* of sense perceptions themselves, reflecting the latter of Eagleton's (1991) two strands. Horkheimer in particular emphasises the way that knowledge reflects the current material conditions of life, or as Linda Martin Alcoff (2007) summarises it rather succinctly: 'knowledge, no less than 'subway trains and tenement houses,' reflects the current condition of human praxis.' (Alcoff, 2007: 54). For Horkheimer and thinkers aligned to the Frankfurt school, to understand the ways that even looking and how our sense faculties are actively shaped for us we must pay attention to the whole arrangement and "praxis"—the practice and theory—of human life today as it presents itself. Ultimately, Horkheimer's point is that we cannot disentangle our perceptions and knowledge of the world from the predominant mode of production and reproduction of life itself. To put a finer point on it, to understand how perceptions are shaped we need to understand the whole system of capitalism itself. This is not, of course, to suggest that having read Marx's *Capital* grants an "ideology-free" perception, but rather that understanding capitalist praxis makes visible the previously invisible ways that what we see and how we know is variously enabled or disabled, supported or resisted, helped and hindered by the different material, political and economic arrangements of the world around us. This, in part, is why the rest of the chapters of this book focuses so heavily on the production of games, the material stuff that gets made, and the economic forces involved: from electrons generated by photonically excited wafers of silicon crystals, to the thick strands of copper windings that spin under steam pressure, even the conditions in the workplaces that where the actual work of art, design and coding that it takes to produce the modern videogame. This is why we must insist on going further than simply stopping at questions of cultural interventions, social 'nudges', consumer shifts to or away from this or that form of consumption.

To return to the question of ideology and interventions—Eagleton argues that anyone attempting ideological critique should take note that 'only those interventions will work which make sense to the mystified subject itself.' (Eagleton, 1991: xiv) This is worth dwelling on, as it underscores the nature of the problem as one that cannot be solved automatically, for instance via the simple presentation of "reality". Games hoping to do so may well be wasting precious time and energy, quite literally. This point is perhaps Eagleton's (1991) most practical contribution, highlighting that ideological struggle involves contests over meaning and other systems of knowledge and entangled with the material conditions of modern life. Just like Horkheimer, Eagleton (1991: xiii) describes ideology—that which 'allows men and women to mistake each other from time to time as gods or vermin'—as deeply rooted in the conditions of contemporary life and social power, and the internal struggles of individuals with their own (contingent and historical) tastes, preferences, and sensibilities. This perspective is also reflected in the latest developments in communication theory, and their application in climate communication. As described by Ballantyne, (2016) the 'interaction paradigm' of communication that has replaced earlier 'information deficit' approaches, going further than simply addressing a perceived lack of knowledge or information about the climate crisis to emphasise that we must 'conceptualize sender and receiver as active participants in a process of co-creating meaning' (Ballantyne, 2016: 339). There are parallels here with what I called for earlier in the literature on climate games—with a more robust and detailed knowledge of the individuals themselves required. Unfortunately, Ballantyne (2016) finds that 'many initiatives within climate change communication are still driven by an objective of addressing a public information deficit' (Ballantyne, 2016: 339). While this situations may be slowly improving, we are still far from where we should be, particularly regarding games attempting climate communication. Despite their oft-emphasised interactive nature, games still face conceptual and practical challenges in achieving the sort of player knowledge that the interaction paradigm of communication suggests—especially at scale. Ballantyne (2016: 339) notes that,

> According to [the interaction paradigm of communication], the way forward for climate change communication research could be (a continuous) exploration of how people constitute the reality of climate change. What does climate change mean to different audiences and why? What causes people to engage or disengage with climate change? How do climate change discourses contribute to the reproduction, maintenance, or transformation of the perceived reality of climate change?

Perhaps conveniently for those games and developers that wish to achieve effective climate communication, Ballantyne (2016: 339) emphasises that such an approach 'does not necessarily disregard strategic communication such as information campaigns. It merely contends that communication cannot be understood correctly unless acknowledged as a constitutive process.' The importance of this constitutive view of communication cannot be overstated, suggesting we need to understand all 'communication as a social process that constitutes reality for the participants of that process' (2016: 340). In this way, we are necessarily guided to recognise in our climate communication (including any communication that involves the use of games) a more complex relationship than simply that between a sender and receiver, or the presentation of 'reality' via a computer model or simulation. It also presents a challenge to game makers who I suspect, even with detailed market knowledge about particular segments of the gaming audience, are unlikely to know exactly how the reality of climate change is constituted for their players. This sort of audience knowledge does not come easily to the mass-market of game development, nor even for those targeting gaming niches. In many cases games simply cost too much to make, and therefore must recoup too much money from too large an audience to care about individuals. Even for non-didactic games and those seeking to do less complex climate messaging, effective communication will be a struggle—not least of all because of the challenges of ideology and the risk of running into the problems of simulation fever. Games that wish to perform ideological critique will need to make sense for the player playing them and offer their climate communication in such a way that fits with their particular social processes of reality construction—which could look drastically different from person to person. It might even, in this way, rule out the very idea of a 'mainstream' climate game, one which would communicate about the climate in a way that makes sense to every player that plays it.

Eagleton (1991: 5) also informs us that often 'ideology has to do with legitimating the power of a dominant social group or class' and this observation extends to both the ideology embedded in—and challenging ideology of players through—games. The multimillion-dollar projects of the Triple-A games industry seem particularly unsuited to embodying a challenge to ruling ideology and dominant social powers, though it may not be impossible. I think it is fair to say that not many games, even the ones that wish to make some kind of ideological critique, currently are designed to take into account the diversity of player's positions within dominant political, economic, material and ideological structures. For climate activists and game designers wishing to use the tools they know best to change the world, the diversity of dispositions of players is an irreducible problem—possibly an insurmountable barrier that simply will not go away under the current arrangement of the games industry (Horkheimer's 'current condition of human praxis'—just like subway trains and tenement houses, so too games are subject to and reflect the same forces). Unless we are willing to accept that the pursuit of ideological persuasion through games is reserved for an entirely artisanal section of the industry, one that produces highly targeted works aimed potentially at an audience of one, tailoring their communication to these prospective players, then the problem of players and their ideological resistances simply cannot be papered over by enthusiasm for what might be or could be possible through game-based activism. Perhaps it's not impossible that such a change could come about. But consider, for example, Nicoll and Keogh's (2019) work interrogating the supposed 'democratising' claims made about the Unity game engine and the persistence of platform capitalist dynamics therein. If we want to argue for mainstream, transformative cultural change through games then we will need to pay attention to those games actually in the mainstream, and the conditions of their production that make them the way they are.

Is it possible for games to perform ideological critique? Despite these barriers, I still think it may be, in rare cases. The next section finds, in a most unlikely game, an optimistic, hopeful climate future created neither through models or simulations but through an aesthetic vision of the climate future—in the military sandbox game *ARMA 3*.

## *ARMA 3*'s Aesthetic Vision of a Climate Future

*ARMA 3* (2013) is a military first-person shooter simulation game, touted by its developers Bohemia Interactive as 'a massive military sandbox.' Launching initially without even a single player campaign, just the tools for fans to create their own scripted scenarios and capacity for modding, the game is primarily a multiplayer sandbox with scope for a range of gameplay types within the limits of the *ARMA 3* engine's infantry and vehicular focus. The base game (there have now been a number of expansions) takes place on one of two large islands, with no set levels, checkpoints, or invisible walls preventing players from moving around, giving the game a sense of existing in one massive connected space. *ARMA 3* is often described as more akin to a "platform" than a game-as-such. YouTube content creator and occasional Bohemia Interactive consultant Andrew "Dslyecxi" Gluck (2013) describes *ARMA 3* as 'the Minecraft of military-sim games.'

Players who spend any significant time with the game will typically discover that the main island on which *ARMA 3* is set, modelled on the terrain of the real Greek island of Lemnos (referred to in game as "Altis"), features numerous renewable energy installations strewn across its 270 km² terrain. An aerial tour of the island reveals wind farms dotted along several prominent hills, with their lazy blades chopping the air, as well as industrial scale solar thermal installations near the island's main airfield focusing the virtual sun's rays onto a central tower to generate steam for an imaginary turbine. Fields of solar photovoltaic panels are dotted around the island, and, if one knows where to look, tidal power generating buoys float just below the surface of the waters off the southern coast, with one or two hauled up on the beach. The whole island is noticeably absent the familiar structures one associates with twentieth-century fossil fuel power generation—no iconic cooling towers of coal or nuclear fired power stations with plumes of rising steam. Instead, there is only the bright, reflective solar panels, futuristic thermal towers, and an impressive abundance of wind farms. The aesthetic of the island of Altis is dominated by the visible presence of renewable power generation.

Set in the near-future of the 2030s, *ARMA 3's* designers devoted much time to evoking a very futuristic feel, particularly through weapon and vehicle designs, felt all the more keenly on its release back in 2013. It introduces camouflage patterns plausibly extrapolated from contemporary pixel-based uniforms, as well as weapons and vehicle technologies that appear as logical projections based on current designs, evoking a plausible, near-future—one which just so happens to feature renewable power technology. In this way *ARMA 3* presents a more compelling and optimistic vision of the future than many other games which set out to do it deliberately.

Ziser and Sze (2007) have noted that Western activism and particularly the visual art world's engagement with climate change has often invoked the feeling of the sublime to impress upon the viewer the massive scale of the issue. They argue, however, that we should reject these appeals, and the overwhelming aesthetic of the sublime, noting instead that 'environmental justice aesthetics ought to reject the sublime scale invoked by some [global climate change] narratives and instead remain focused on the human, ecological, and social justice dimensions of environmental change' (Ziser and Sze, 2007: 407). Their argument indirectly reflects a sensitivity to the problem of simulation fever—seeing a model or simulation and rejecting its implications—highlighting that sublime depictions of climate change can be paralysing. Perhaps counterintuitively, *ARMA 3* makes no attempt to model or simulate actual power generation in its world. Nor does it reproduce the process required to get from "here" to "there" politically or economically in the struggle to decarbonize the world. Nor does *ARMA 3* model the resistance to change and the political lobbying of the huge vested interests arrayed against just such a transition. It does not model the sensational and spectacular impacts of climate change—the background of increasing risk of devastating climactic events, rising sea levels, permanent changes to environments, and so on. No mechanics or appreciable "gameplay" is even directly involved with any of these issues. Instead, it simply presents the player with the visual and aesthetic indications of renewable power generation, with no added textual or narrative explanation or context. It is even quite likely that featuring renewable power generation methods in the game was not even done to make a political point—just as likely they are present simply as part of evoking the feel of a near-future technological era.

The effect, however, is in stark contrast to the earlier approaches and the literature which sought to find merit in engaging with the crisis through simulations, or mechanics, or even 'inhabiting' conscious scenarios of resource scarcity. The approach of *ARMA 3* and its aesthetic vision of the future is almost the reverse, to recede into the background, presenting a vision of the near-future purely aesthetically, via 3D rendered space already pre-populated with renewables. There is precious little space left for conscious rejection of the scenario precisely because it is so very thinly instantiated. *ARMA 3* thus demonstrates that other approaches are possible, and that engaging the issue indirectly, in a roundabout way is possible—one which largely does not leave room for the conscious interrogation or rejection of climate change mechanics. As with all things regarding the living, breathing player, however, there is still no guarantee that the desired effect will be achieved. But it at least seems plausible that aesthetic approaches might be able to avoid some of the conscious barriers and resistances outlined earlier.

*ARMA 3* offers the player the opportunity to perform what Eagleton (1988) describes as the "self-delight" of social submission to an aesthetic regime. This submission consists of simply inhabiting a vision of the future, and likely happens below the level of conscious acceptance or rejection (unlike intellectual engagement with mechanics). It suggests a more seductive approach, one which appeals to a different aspect of a player's inner life and sense-making, and a potential detour from simulation fever's conscious rejection of mental models that clash with one's ideological commitments. *ARMA 3's* aesthetic depiction also has the advantage of avoiding the scale of the sublime, which may cause players to turn away from the problem, overwhelmed. I find an aesthetic engagement, the use of art style, visual design, and the use of near-future references that speak to climate solutions like the renewable technologies found in *ARMA 3* results in a rather subtle and sophisticated engagement with the highly ideological issue of climate change. This, to me, is where games might have more potential to be much effective, and stand a greater chance of making some sort of enduring cultural impact—not by the conventional attempts to persuade players in a didactic manner through mechanics, simulations, and the like. How do you argue with a picture? How do you reject an impression, a sense, or a feeling? To be sure, players may still very well reject this—there is no guarantee that players take what we want them to, as Gee (2003) noted. Even the most appealing description or imagining of a future filled with plentiful renewable energy may be dismissed, rejected as 'silly' or 'unrealistic'. But there seem to be fewer barriers, and in this case, it also has the distinct advantage of being a game that already appeals to a certain type of gamer. This is not to say that military sim players are climate deniers, but that they are more likely (I suspect) to be playing *ARMA 3* than they are a game that is consciously about climate change. Climate games that want to perform ideological critique must, at the very least, appeal to, or makes sense for the mystified individual themselves. Given what we know about the necessity of active participation in learning, and that climate communication needs to makes sense within the social processes of knowledge making in the person's life, it is hard to see how the ideologue could be convinced by any number of simulations or representations of real-world climate or environmental dynamics, no matter how great its fidelity to the world itself.

But there may be alternatives to this kind of direct assault on the "mystified" player's mental models and beliefs, and perhaps we are able to do so while keeping in mind both the importance of the social processes involved in sense-making from communication (Ballantyne, 2016), while also acknowledging the historically and culturally conditioned sense data (Horkheimer's (1972) notion of human praxis). *ARMA 3* says aesthetically that a world of renewable energy is possible, and crucially it does not say this in words or through the traditionally prioritised pipeline of meaning-making in games via mechanics and design, but through a seductive appeal to a nicer, more renewable world.

## Conclusion

This chapter's primary goal has been to examine whether, and if so how, games might save the world and help avert climate crisis. Beginning with McGonigal's (2011) proposal that all the broken aspects of the world could be fixed by games, I outlined what I see as the shortcomings of a project largely aimed at rectifying the failures of individuals. This pattern continued in the literature on green games and their supposed powers for persuasion and change which left an absence in the figure of the player. For most of this work, the actual end-user or player—the living human being in all their complexity and the local situated context they play games in—is largely glossed over. The lack of such pesky details as race, gender, class and a host of diverse social processes of meaning-making enables such enthusiastic accounts of games 'potential' (rarely actualised) for this or that intervention. Turning to the literature on games, learning and persuasion that does account for more of the lived context of players, the chapter then identified further barriers that exist to instrumentalist deployments of games for climate or environmental ends. One of the most substantial of these barriers is ideology, the operation of which is rarely addressed. Ideology is especially problematic for games seeking to instil knowledge, or promote learning through player reflection on their experience with mechanics, dynamic models or simulations. The threat of simulation fever—rejecting the contours and outcomes of a simulation—can never be fully eliminated, and this undermines accounts that ignore this possibility. Finally, I found in the military-simulation game *ARMA 3* an alternative approach in the form of a more seductive appeal to the senses of the mystified player, one that does not need to engage rational or conscious deliberative faculties, giving them the opportunity to 'submit' to an aesthetic regime of a future filled with renewables. As with all accounts of what a game might do there is still no guarantee of success, but at least it represents a plausible way overcome the ideological walls thrown up around the issue by the mystified player.

What I have been arguing for throughout this chapter is the necessity of a serious account of players themselves in any work that seeks to engage with or articulate the powers of games to intervene on a given player. This means being cognisant of the fact that a process of socially constructing meaning is occurring whenever even the most detailed or faithful simulation of a climate or environmental dynamic is encountered. At an absolute minimum, accounts of games 'powers for persuasion' should speak of the 'willing player'—meant both in the sense of being willing to go along with or accept the climate communication, environmental lesson, or the simulated model, as well as in the sense of being a wilful subject. This would at least tacitly acknowledge the equally *unwilling* player in the same moment.

It seems to me that the problem that would-be games for activism face is the same problem facing climate activists all over the world. For both, the question remains: "how do you make someone care about the planet?" To some this seems like a simple problem of pedagogy, of instruction, or the (now mostly defunct) 'information deficit' model of communication. However, I want to affirm a different model entirely, one that Jacques Rancière in *The Ignorant Schoolmaster* outlines, along the lines of a principle of radical equality. Rancière explains:

> There is no one on earth who hasn't learned something by himself and without a master explicator... "Everyone has done this... a thousand times in his life, and yet it has never occurred to someone to say to someone else: I've learned many things without explanation, I think you can too."
> (Rancière, 1991: 16)

Rancière's radical proposal is based on an approach by Joseph Jacotot in the nineteenth century, that simply believes anyone can learn anything at all they want to, that we all have equal intelligence, equal capacity to come to an understanding, to learn what others have learned, or even to learn nothing at all! For Rancière this approach is important as it avoids what he describes as the stultification of instruction, a kind of 'pedagogical myth... [that] divides the world in two. More precisely, it divides intelligence in two. It says there is an inferior one and a superior one' (Rancière, 1991: 7). For Rancière (1991) the explanation of one who knows to one who does not yet know—even done in good faith and with the best intentions in the world, even we might say, in an attempt to impart the knowledge to care and respect for the very earth that sustains and gives life—results only in 'stultification' and the subordination of the learner's own intelligence. Education is not what climate sceptics and those with insufficient love for the earth needed, but emancipation, being set free:

> Whoever teaches without emancipating stultifies. And whoever emancipates doesn't have to worry about what the emancipated person learns. He will learn what he wants, nothing maybe. He will know that he can learn because the same intelligence is at work in all the productions of the human mind.
> (Rancière, 1991: 18)

If the homogenous 'player' that lurks undefined behind so much of the literature on games 'power to change' has any affirmative characteristics at all, then it involves an assumption of lesser intelligence. It presumes a learner needing to be taught. Simply posing the question "can games teach climate change?" presupposes such a player whose mind needs this or that instruction, this or that correction, intervention in some way. If there is one thing I would try to impress upon the reader of this chapter it is the importance of throwing away this notion that players "need" anything from us at all (anything, perhaps, except emancipation). This might seem like it's of no help to climate activists, but on the contrary, it helps us immensely by clarifying the nature of the task. It redirects our effort from instruction to emancipation, and opens up an entirely new domain of questions.

Another way to phrase our application of Rancière's insight might be to say that in order to get players to care about the planet, they first have to *want* to care about the planet. This might seem a pointless tautology, or a splitting of semantic hairs, but actually it refocuses our attention on the will, the desire, the wants and needs of a person. These are desires that are socially and historically 'preformed'—and could be, or could come to be otherwise:

> The facts which our senses present to us are socially preformed in two ways: through the historical character of the object perceived and through the historical character of the perceiving organ. Both are not simply natural; they are shaped by human activity. (Horkheimer, 1972: 200)

This is what Horkheimer's expansive phrase the 'current condition of human praxis' encompasses—the awareness that our desires don't just spring unbidden from a hidden internal reservoir, they are formed by the very historical and material processes at work in the world around us, by "subways and tenement houses" and these days as much as by the video-game industry and its political-economic consequences. Cultural critic Mark Fisher and Matt Colquhoun (2021) recognised this keenly and emphasised it throughout his work, highlighting the importance of elucidating and cultivating what he called a 'post-capitalist desire', the title of his final, unfinished lecture series. Pelletier (2012) also points out, in her analysis of Rancière's relationship to modern pedagogical theory, that it's possible to "avoiding equating emancipation with freedom from [all] constraint (i.e. 'whoever wants is able to')" (Pelletier, 2012: 103). The emancipated player is not a totally free and unfettered will, in other words. There are real constraints on people's 'wanting' to care about the planet. Hostile media ecosystems, class interests, family traditions, and social norms all bear down upon us in different ways—but a focus on where and how these wants and desires are produced can help us to target our work at the various centres of power and influence in the social whole, and avoid stultifying impulses to correct the masses into right-thinking, green-conscious consumers. Farca, (2016) also drawing on an essay of Rancière's (2009) argues for a notion of the 'emancipated player' that has real resonances with what I am calling for here, though a slightly different emphasis. For Farca (2016: 3) this is a particular player who 'frees herself from a confining and linear perception of video games and acknowledges their multifaceted nature.' The emancipated player, is not a given, and cannot be assumed, but rather exists as potential that 'slumbers in all of us' (Farca, 2016: 8). According to Lehner's (2017) reading of Farca's concept, the emancipated player's full engagement with the aesthetic object of the game renders redundant notions that any one part or feature of a game is primary: 'for example Bogost's (2007) structural notion about the primary importance of the procedural rhetoric of videogames' (Lehner, 2017: 58). The emancipated player does not just pay attention to mechanics and rules, but the entire spectrum of possibilities of intellectual and aesthetic engagement with a game.

What I am arguing for then, is actually something much bigger in scope than changing the world through games and the people who play them. As I have noted, I don't hold much faith in that approach in any case. What we need is to change the world itself—the world that players inhabit; the world that 'preforms' their senses, that provides them with the situated context in which they live their lives.

What sort of changes? It is the task of the rest of this book to outline some of the possibilities, and readers may well have ideas beyond what I can describe. But it is precisely in the world itself that change is most needed—in the world of shit, dirt and trash, as Chang (2019: 173) so evocatively put it. That is where we need to act. If we can change the emissions footprint of the games industry itself, as I propose that we must, we render the mystified player's objections almost irrelevant—we have made a better world regardless.

We have decades left. Maybe even less. Games, if they are to be a force for change, like any other cultural and artistic endeavour, are not going to play the role of teacher in any meaningful pedagogical sense—at least, not with the games industry organised and structured the way it is currently. They cannot be relied upon to change either willing or unwilling players— not with any certainty or any guarantee of success. We do not have the luxury of outlandish diversions anymore. Some may want to extrapolate from the claim I made at the outset of this book that the problem is now so urgent, so dire that we must do everything we can—but an important part of that sober assessment is recognising we are finite ourselves. We do not have the luxury of continuing to try things that have been ineffective thus far. We will need to prioritise some things over others.

There are more important actions to take right now, with more direct impact on emissions—as we shall see in the latter half of this book—than trying to convince or improve an unspecified Player. Beyond a small num- ber of examples, the instances of games making deliberate, targeted inter- ventions in players opinions have mostly been unsuccessful or fleeting—especially so on the scale that the crisis demands. Games do not perform the same function as social movements, political-economic revo- lutions, or even policy prescriptions—they are just games! They are often distractions, a hugely entertaining way of spending time. These are on their own worthwhile and in no need of buttressing with additional social objectives. To the degree that any particular game shifts public conversa- tions, they are almost always in trivial or at best unpredictable ways. Which games become the next cultural juggernauts like *Minecraft*, *Fortnite*, and so on is impossible to know in advance—how much more so games for concerted social impact? We don't expect other cultural forms to carry the baggage of such grand ambitions as changing society—ballet, opera, film rarely (if ever) act on the popular consciousness or even the individual mind in predictable behaviourist ways. All these cultural forms and more are worthy in their own right, and we do games a disservice when we attempt to overload them in this way. For all their incredible importance to contemporary pop culture, games do not have a terribly important place in most of our lives—even those of us who play a lot of them. They don't typically change how we live our lives, what we feel in our hearts, or attack our core beliefs and convictions. This should be an uncontroversial observation, but years of defensiveness has built an edifice around games to shield them from the slings and arrows of outsider perceptions and accusations.

When it comes to games, then, we should release ourselves from the ultimately stifling and unsatisfying attempts to identify any clear and cat- egorical effects of games *tout court*, and breathe a sigh of relief as we do so. Given all that I have said, it seems almost impossible to reliably identify let alone design for uniform or predictable effects on players *en masse*. Crucially, this does not entail writing off identifying the effects of particular games on particular players—especially given the plenitude of stories, anecdotes, after action reports, and a host of other accounts that players themselves frequently share. The kind of storytelling and anecdote swapping amongst friends or, just as commonly among community, which takes the form of personal writing, criticism, and even games journalism only serves to underscore the complexity regarding the "effects" of any form of communication. The non-uniformity of these experiences, even as they remain recognisable from one player to another, underscores the importance of the social process of understanding in the interaction between player and game, one which is embedded in the wider field of social, political and economic production. Even if we were to find one or more persons affirmatively 'persuaded' by a given climate game we cannot work 'backwards' to deduce from that a strong position on games as a whole and their persuasive and interventionist potential. Players as a "whole" do not exist, certainly not as a lumpen mass waiting to receive the imprint of the game and their meaning, message, or communication. Players are themselves (much like the world itself) too recalcitrant, too diverse, too unpredictable, too playful, confounding the best laid plans of the most effective game designers. We do both games and their players a disservice when we allow ourselves to get caught up in sweeping gestures that attempt to encapsulate the power or potential of the medium, as compelling as these can be.

The remaining chapters of this book aim to show that there are some very practical, concrete and achievable changes that can be made in and around games *today* which are almost guaranteed to have substantial positive impacts. None of these changes depend on a willing player subject, and none of them face the ideological barriers of the mystified player. Instead they take aim at the world these players exist in, both the willing and the recalcitrant, and aim to change them for the better, with or without them. By reducing the impact and emissions of the games industry itself, the views of players become almost irrelevant. The next chapter asks what a *truly* ecological game would look like, arguing that it must be one that pays much more attention to the conditions of both its own production and the energy and emissions context in which it is played. The remainder of the book and the chapters that follow, turns our attention towards the very real and very substantial carbon costs involved in maintaining the games industry, and how companies and individuals can start to take important action to decarbonise today. Inevitably, these involve rather less exciting but far more critical political and economic issues around the makeup of what is still, in fact, a substantially carbon intense, and highly unsustainable industry.

## REFERENCES

Alcoff, Linda Martín. (2007). "Epistemologies of ignorance: Three types" in Sullivan, Shannon and Tuana, Nancy (eds.) Race and Epistemologies of Ignorance. Ithaca: State University of New York Press.

Apperley, T. (2011). Gaming rhythms: Play and counterplay from the situated to the global. Institute of Network Cultures.

Apperley, T., & Jayemane, D. (2012). Game studies' material turn. Westminster Papers in Communication and Culture, 9(1), 5–25.

Ballantyne, A. G. (2016). Climate change communication: What can we learn from communication theory? Wiley Interdisciplinary Reviews: Climate Change, 7(3), 329–344.

Barton, M. (2008). How's the weather: Simulating weather in virtual environments. Game Studies, 8(1), 177–193.

Bell-Gawne, K., Stenerson, M., Shapiro, B., & Squire, K. (2013). Meaningful play: The intersection of video games and environmental policy. World Future Review, 5(3), 244–250.

Bogost, I. (2007). Persuasive games. MIT Press.

Bohunicky, K. M. (2014). Ecocomposition: Writing ecologies in digital games. Green Letters, 18(3), 221–235.

Chang, A. Y. (2011). Games as environmental texts. Qui Parle: Critical Humanities and Social Sciences, 19(2), 56–84.

Chang, A. Y. (2019). Playing nature: Ecology in video games. University of Minnesota Press.

Chess, S. (2017). Ready player two: Women gamers and designed identity. University of Minnesota Press.

de Suarez, J. M., Suarez, P., Bachofen, C., Fortugno, N., Goentzel, J., Gonçalves, P., Grist, N., et al. (2012). Games for a new climate: Experiencing the complexity of future risks. Pardee Center Task Force Report.

Dyer-Witheford, N., & De Peuter, G. (2009). Games of empire: Global capitalism and video games. University of Minnesota Press.

Eagleton, T. (1988). The ideology of the aesthetic. Poetics Today, 9(2), 327–338.

Eagleton, T. (1991). Ideology: An introduction. Verso.

Elson, M., & Ferguson, C. J. (2014). Twenty-five years of research on violence in digital games and aggression. *European Psychologist, 19*, 33–46. https://doi.org/10.1027/1016-9040/a000147

Farca, G. (2016, August). The emancipated player. In *Proceedings of the joint DiGRA & FDG conference*. DiGRA and Society for the Advancement of the Science of Digital Games.

Fisher, M., & Matt, C. (2021). Postcapitalist desire : the final lectures. London: Repeater Books.

Galeote, D. F., Rajanen, M., Rajanen, D., Legaki, N.-Z., Langley, D. J., & Hamari, J. (2021). *Gamification for climate change engagement: Review of corpus and future agenda*. Environmental Research Letters.

Gee, J. P. (2003). What video games have to teach us about learning and literacy. *Palgrave Macmillan, 14*, 203–210.

Gluck, A. (2013). Intro to Arma 3 - Dslyecxi's Arma 3 Guides. YouTube. Retrieved March14, 2013, from https://www.youtube.com/watch?v1⁄42XdK9xhJyH8

Guins, R. (2014). Game after : a cultural study of video game afterlife. Cambridge, Massachusetts: MIT Press.

Haraway, D. (1991). *Simians, cyborgs, and women: The reinvention of nature*. Routledge.

Horkheimer, M. (1947). Eclipse of reason. *Continuum*, 1974/2004.

Horkheimer, M. (1972). *Critical theory: Selected essays*. Continuum.

Huber, M. T. (2013). *Lifeblood: Oil, freedom, and the forces of capital*. University of Minnesota Press.

Kelly, S., & Nardi, B. (2014). *Playing with sustainability: Using video games to simulate futures of scarcity*. First Monday.

Latour, B. (1993). *The Pasteurization of France* (A. Sheridan., & J. Law, Trans.). Harvard University Press.

Latour, B., & Woolgar, S. (1986). *Laboratory life: The construction of scientific facts*. Princeton University Press.

Lehner, A. (2017). *Videogames as cultural ecology: "Flower" and "shadow of the colossus"*. Ecozon@.

MacKenzie, D., & Wajcman, J. (Eds.). (1985). *The social shaping of technology: How the refrigerator got its hum*. The Open University Press.

Malm, A. (2016). *Fossil capital: The rise of steam power and the roots of global warming*. Verso Books.

McGonigal, J. (2011). *Reality is broken: Why games make us better and how they can change the world*. Penguin.

Meillassoux, Q. (2008). *After finitude: An essay on the necessity of contingency*. Continuum.

Moore, J. W. (2015). *Capitalism in the web of life: Ecology and the accumulation of capital*. Verso Books.

Newman, J. (2012). Best before : videogames, supersession and obsolescence. Milton Park, Abingdon, Oxon New York: Routledge.

Nicoll, Benjamin, and Brendan Keogh. The Unity game engine and the circuits of cultural software. Cham, Switzerland: Palgrave Macmillan, 2019

Pelletier, C. (2012). No time or place for universal teaching: The ignorant schoolmaster and contemporary work on pedagogy. In J.-P. Deranty & A. Ross (Eds.), *Jacques Rancière and the contemporary scene* (pp. 99–116). Continuum.

Raftopoulos, M. (2020, April 1–3). Has gamification failed, or failed to evolve? Lessons from the frontline in information systems applications. In *GamiFIN* (pp. 21–30). Levi.

Rajanen, D., & Rajanen, M. (2019). Climate change gamification: A literature review. In *GamiFIN* (pp. 253–264).

Rancière, J. (1991). *The ignorant schoolmaster*. Stanford University Press.

Rancière, J. (2009). *The emancipated spectator*. Verso.

Reckien, D., & Eisenack, K. (2013). Climate change gaming on board and screen: A review. *Simulation & Gaming, 44*(2–3), 253–271.

Salen Tekinbaş, K. (Ed.). (2008). *The ecology of games: Connecting youth, games, and learning*. MIT Press.

Stevens, R., Satwicz, T., & McCarthy, L. (2008). In-game, in-room, in-world: Reconnecting video game play to the rest of kids' lives. In K. Salen Tekinbaş (Ed.), *The ecology of games: Connecting youth, games, and learning*. MIT Press.

Szycik, G. R., Mohammadi, B., Münte, T. F., & Te Wildt, B. T. (2017). Lack of evidence that neural empathic responses are blunted in excessive users of violent video games: An fMRI study. *Frontiers in Psychology, 8*, 174.

Winner, L. (1986). Do artifacts have politics? In *The whale and the reactor: A search for limits in an age of high technology* (pp. 19–39). University of Chicago Press.

Wu, J. S., & Lee, J. J. (2015). Climate change games as tools for education and engagement. *Nature Climate Change, 5*(5), 413–418.

Yusoff, K. (2018). *A billion black Anthropocenes or none*. University of Minnesota Press.

Ziser, M., & Zse, J. (2007). Climate change, environmental aesthetics, and global environ-mental justice cultural studies. Discourse, *29*, 2–3.

# CHAPTER 3

# What Is an Ecological Game?

In the previous chapter, I concluded with the challenging proposal that climate activists face substantial barriers to persuading players via game experiences. These barriers are especially pronounced when faced with an ideologically mystified player, as may very well be the case in climate communication and where such communication 'making sense' to the receiver is essential (Ballantyne, 2016). This challenge of personalization of communication to specific individuals also seems mismatched with games that aim to reach a mass audience. Unless we are willing to accept game-based climate communication remaining niche, or a cottage industry, and thus give up on taking advantage of games popular appeal—then we may need to seek other avenues for effective climate action.

What if we approach this problem from a different angle, however, shifting our focus onto the game object itself? What if we assume there may be some other benefit to green games beyond the persuasion of climate sceptics or the instruction of individuals? Might there be something to be gained by games representing the environment for their own sake, even if we cannot guarantee any sort of predictable change or outcome in the players who play them? There is certainly no shortage of both scholars and developers who see some value in making and playing these kinds of games, and in any case it's unlikely that we will see the end of them, in which case, perhaps they can be improved. Is it possible, for instance, to embody specific environmental principles and ideals in a game, or make

© The Author(s), under exclusive license to Springer Nature
Switzerland AG 2022
B. J. Abraham, *Digital Games After Climate Change*, Palgrave
Studies in Media and Environmental Communication,
https://doi.org/10.1007/978-3-030-91705-0_3

games so that, as Alenda Chang (2019: 11) suggests, they become 'boundary objects that facilitate passage between the material and seemingly immaterial contexts of the physical world and virtual playspace?' What I am really asking here is *what is an ecological game?*

To answer this question, the chapter begins with a brief discussion of the ecological and environmental metaphors and discursive resonances that suffuse wider game discourse, both popular and scholarly, to establish what is considered 'ecological' or environmental about games at present. These tendencies find perhaps their greatest expression in the genre loosely defined as survival-crafting games, with games like *Minecraft* (Mojang 2011) which have been widely regarded as possess mechanical and structural features that resemble or align them more closely to real environments. Analysis of the survival–crafting genre's two principal distinguishing features however—namely, survival and crafting mechanics—does not reveal a particular ecological ethos, however, and the structural features common to these games (those that receive the lion's share of attention from players, press, scholars and designers) actually bears far more resemblance to the structures and organizing logics of technologically advanced capitalism. I argue that even if pro-ecological or counter-readings of these types of games are *possible*, their underlying logic resembles or reimposes most, if not all, of the same problems most in need of dismantling. The "survival" aspects of most of these games guide players toward dynamics of accumulation and affective experiences of life common to capitalism, like eating rendered a burden or nuisance, or the problem of maximally efficient use of one's time. In the other half of survival–crafting, in crafting systems and mechanics we frequently see the reproduction of the ahistorical trope of the ladder of technological progress, coupled with representations of human-environmental relationships that resemble and reproduce problematic narratives of mastery and domination of the natural world.

In response, the chapter concludes by arguing that the truly ecological game must acknowledge and account for the harms it produces in the world, its own direct and indirect material costs for the environment, and its participation in worsening (or aiding) the climate crisis. The truly ecological game, in this way, achieves the goal, established in the previous chapter, of rendering players' ideological resistance to climate action redundant through directly changing the material world the player themselves inhabits, and participates in a wider political project of social and economic transformation.

## ECOLOGICAL RESONANCES IN GAME
## DISCOURSE METAPHORS

One hardly has to look hard for big–budget, Triple-A mainstream games that employ design elements and techniques that were, until recently, mostly relegated to specific genres like the "simulation" or "sandbox" games. Huge, industry—shaping titles like *No Man's Sky, The Elder Scrolls V: Skyrim*, and the *Far Cry* Crytek, Ubisoft Montreal, Ubisoft Toronto 2004–2021) series make much use of design tropes that have been described by developers, critics, and scholars alike as "emergent"— with the story or narrative "emerging", at least in part, in a dynamic, unpredetermined fashion out of the complex interactions of a host of different systems and objects. In *Skyrim (Bethesda Game Studios 2011)*, wild animals, humanoids of different warring factions, and the seemingly random appearance of dragons, all clash together violently and unpre- dictably in the wide-open spaces of the game. The interaction of an extensive network of mostly invisible code and scripting systems generat- ing many possible 'narratives' and events, from heroic battles of humans vs spiders, bandit raids on caravans of travelers, all of which can be inter- rupted by the appearance of a third party like trolls or other powerful monsters, and any number of other strange or unlikely occurrences. All sorts of micro-encounters "emerge" seemingly unbidden out of the net- work of systems and semi-randomness.

The concept of emergence has been a key one in games for a long time. In his formative essay "Game Design as Narrative Architecture," Henry Jenkins (2004) describes emergent narratives in the popular game *The Sims*, sketching out the function of systemic story elements in games: "Emergent narratives are not prestructured or preprogrammed, taking shape through the game play, yet they are not as unstructured, chaotic, and frustrating as life itself." In Jenkins' (2004) description there is a sense in which emergent games reflect something about the generative nature of the world itself (though incompletely), by producing situations or sce- narios in excess of those determined in advance. Players and critics often seem to prize these elements of emergent games for delighting and con- founding expectations—like *Skyrim*'s pitched battles of unpredictable foes—an ideal often strived for in similar games. In the second edition of Ernest Adams' (2009) *Fundamentals of Game Design* textbook, he pin- points the introduction of the term 'emergence', describing a lecture given by Marc LeBlanc at the Game Developers Conference in 2000 as the place the term was first raised. Adams (2009) devotes only a few paragraphs of description to the entire topic in this edition of his textbook, downplaying the conceptual complexity contained in the term and the prominence it would come to play in game design trends. In 2013, Warren Spector, one of the principal developers responsible for the *Deus Ex* (Ion Storm 2000) series of games, the first title widely credited with showing the potential and sparking mainstream interest in emergent game systems, offered his "favorite definition of the term [as] 'engines of perpetual novelty'" (Alexander, 2013). Describing how one goes about designing for "emergence" in games, journalist Leigh Alexander paraphrases Spector's advice as follows:

> Create global rules versus specific, instanced behavior of objects and characters; build interlocking systems that are predictable and consistent (some objects are flammable, some guards are lightsensitive, the player has torches) but not predetermined. Have a variety of object properties with plausible or simulated effects ('let water be water') that players can learn and engage with.

This sense of surprise, of generativeness, and of a general indeterminacy coupled with players' own expectations about the behavior of natural objects and systems—for instance, mimicking aspects of the world itself and its operation ("let water be water")—is an attempt to reflect the world and reproduce our complex relation to its confounding qualities of both dynamism and predictability. Player expectations of how water or fire behaves, for instance, are relied upon, in the process bringing with them much of the same cultural baggage that the West is burdened with regarding nature and natural objects—a whole complex of ideas around the repeatability and replicability of "natural" events and behaviours. The prevalence and importance of the term "realism" in gaming discourses reflects a similar preoccupation, despite being greatly dependent on what Alexander Galloway (2004) describes as a form of "social realism." For games, he notes:

> "realism" is, however, a particularly unstable concept owing to its simultaneous, yet incompatible, aesthetic and epistemological claims, as the two terms of the slogan, "representation of reality," suggest.

A full excavation of the depths of meaning associated with the term realism in games discourse is well beyond the scope of this chapter (see, however, Crisp (2015) and Shinkle (2020) for two examples) but for our purposes the important thing to note is the goal of achieving correspondence (via aesthetic and epistemological claims, as Galloway (2004) notes) to an object or behavior that we find, or perhaps more accurately, *expect* to find in or to belong to the world itself. For fire to be deemed "realistic," it needs to behave in certain ways, principally by burning, but at other times by being extinguishable by water. In an emergent game, when fire burns grassland, causing a barrel to explode, killing one's enemies (or allies), the result is often an unplanned scenario "emerging" from the interaction of systemic rules and behaviors. We implicitly understand that this intrusion into our plans by 'natural' behaviors like fire burning was not explicitly scripted by the designer but something that 'emerged' out of the latent possibilities of the game world itself, its rules, its objects, and the programming that determines their interaction. In other words, the emergent scenario is the product (or replication) of an expected dynamic within the natural world itself. There is a kind of correspondence in the emergent scenario that codifies the (usually physical) laws of the universe into a game's system—an attempt to replicate or reproduce "natural" behaviors, and some measure of the predictability of those natural systems. This may seem unproblematic for straightforward phenomenon like fire and water, but when the same schema is applied to other, more complex objects and entities—whether plants, animals, or other living beings with their own agency—the deterministic or systematic nature frequently becomes a problem. If the concept of emergence contains within it a core, metaphoric claim of correspondence between the game and the world itself, the metaphor freights in (intentionally or otherwise) a host of related senses including an affinity with the "natural" or the "real", along with embedded assumptions about what counts or is important to this correspondence.

An even more explicit environmental reference is found in the term environmental storytelling, the emblematic example being the skeleton arranged in such a way as to gesture to the circumstances of its own demise (Spokes, 2018). The *Fallout* series of games is one of the premier culprits of this, with skeletons found on toilets, in bathtubs, side-by-side in a final loving embrace, and in many other arrangements that evoke a sequence of events and a history to these objects through the final moments of their lives. Skeletons are by no means the extent of environmental storytelling however. Jenkins (2004) cites Carson's description which makes this clear:

> Staged areas… [can] lead the game player to come to their own conclusions about a previous event or to suggest a potential danger just ahead. Some

examples include...doors that have been broken open, traces of a recent explosion, a crashed vehicle, a piano dropped from a great height, charred remains of a fire.

Similar to the concept of emergence, environmental storytelling evokes the sense of a living, active world with a prior history or backstory that predates the current game world or the state that it is in. But if this sense of history is evoked rather than instantiated, which is to say it appears to players through elaborate fictional cues rather than being something that *actually happened*, this suggests a fairly thin and awkward conception of the environment, like the two-dimensional background cut-out of a stage performance. It holds up only until you actually look behind it. This raises the question of what actually constitutes an environment in games? And in what sense is it possible for games to have an environment or for that matter to reproduce nature? American philosopher William James once famously claimed that 'nature is but a name for excess' (James, 1909: 63) and game discursive notions that gesture towards the excessive capacities of gaming seem to be reaching for something similar.

Whether it is through emergence or environmental storytelling and the evocation of an imaginary history to a digital space, game discourse is often found repeating these same gestures, reaching for ecological, environmental, or other natural metaphors to explain various systems and dynamics. In the next section, I describe the twin features of survival and crafting that form the core feature of the survival–crafting genre, discussing standout examples and the way mechanics and dynamics in these games sit uneasily with ecological philosophy.

## FEATURES AND DYNAMICS
## OF THE SURVIVAL-CRAFTING GENRE

One of the most successful game genres of the past few years has been given the moniker of survival–crafting, arising in tandem with the increased emphasis on environments, ecosystems, and systemic relationships suggested by the terms and metaphors just described. The two main dynamics common to these games are obstacles to survival which extend beyond the usual adversarial relationship between player and enemy, as well as elements of crafting—whether objects, tools, or creative acts of self-expression upon the land itself.

The prototype for the genre is the cultural juggernaut *Minecraft*, which has also done the most to raise the profile of the genre. Spawning countless imitations and variations on the same design tropes *Minecraft* introduced, it remains something like the pattern from which later games departed. *Minecraft* is often described in environmental or ecological terms, with Amanda Phillips calling the game an "algorithmic ecology" in which 'automated computational processes govern nature… and in many ways, thanks to aesthetic design and game mechanics, subsume the ecological within the mathematical' (Philips, 2014: 109). Similarly, Duncan (2011) describes 'the intricate landscapes and biomes of one's *Minecraft* world', employing the game's own terms for its different landscape regions. Hobbs et al. (2019: 143) claim that 'the *Minecraft* worlds within which each player operates emulate ecologically realistic environments and physical processes.' Ekaputra et al. (2013: 237) even go so far as to claim that 'with functioning ecology, chemistry and physics aspects integrated within the game, these aspects can be used as development media for scientific concept to be used as learning tool for players.' They further underscore the importance of its reproduction of the natural world, saying that '*Minecraft* can be used as fundamental teaching tools for ecology and geology due to its nature [and resemblance to the] real world environment' (Ekaputra et al., 2013: 239). One high-school aged participant in Adams (2012: 10) study involving the use of *Minecraft* in the classroom for teaching seemed to agree, stating that 'in Minecraft you are actually making the farm like an actual farm rather than drawing it on a piece of paper.' Nguyen (2016: 484) likewise notes that,

> *Minecraft* and its two primary playing modes – Creative and Survival – enable players to experiment with various environmental and sociopolitical conditions.

In a particularly detailed analysis of the game, which also catalogues many real world instances of educators using *Minecraft* in schools to teach various environmental concepts, Lane and Yi (2017: 147) describe a litany of mechanics and features that resemble real world environmental or ecological dynamics: from 'exploring and investigating different biomes and climates that match those on Earth, including deserts, forests, jungles, and taigas… [to] interacting with a wide variety of wildlife and agricultural content, including animals, fish, birds, wheat, grass, fruits, vegetables, and diverse fictional content.' Lane and Yi (2017: 149) explicitly highlight a correspondence to real laws of physics in certain mechanics, 'such as grav-ity and flow' in liquids like water and lava. However not all agree—Sharp (2017: 14) perhaps sums this up, describing the geological concepts rep-resented in *Minecraft* as "simplified".

One can easily list a dozen or more games in recent years that have fol-lowed in *Minecraft*'s footsteps in one way or another, emphasizing slightly different elements of environmental dynamics in each: *Terraria, Miasmata, Rust, 7 Days to Die, The Forest, Subnautica, ARK: Survival Evolved*. Games in the genre often feature dynamic, even lush natural landscapes, as well as mechanics that emphasize the player's precarious existence within and dependence upon the natural world. Food and water mechanics are common staples underpinning this precarity, and they contribute to the seeming paradox of an avatar with fairly banal capacities: collecting, chop-ping, building, etc. This is quite different from the power fantasy goals of many other videogames. *ARK (Studio Wildcard 2017)* even takes the uncommon step of modeling regular defecation in humans and animals, including the player themselves, who can use their own feces to fertilize crops (Skrebels, 2016). The marketing material for these games frequently emphasizes the environment as a key element of gameplay, whether laid out by procedural design (*Minecraft, Terraria*) or by hand (*Subnautica, ARK*). Some proceduralization of natural systems occurs in just about all of these games, however, if only at smaller scales such as where and when flora and fauna appear. In *ARK*, for instance, dinosaurs spawn according to algorithms approximating geographic distribution of animal populations.

The 2D survival–crafting game, *Terraria (Re-Logic 2011)*, is described on its Steam online store page with the following zealous exhortation: 'Dig, fight, explore, build! Nothing is impossible in this action–packed adventure game.' Similarly, *The Long Dark (Hinterland Studio 2017)* is described as 'a thoughtful, exploration–survival experience that challenges solo players to think for themselves as they explore an expansive frozen wilderness.' When the game begins, the player is greeted with a disclaimer that notes the developers of *The Long Dark* have taken liberties with ani-mal behavior, in particular its aggressive representation of wolves, noting that they 'are not trying to create 'realistic' wildlife behavior in the game.' The disclaimer also contains the following passage, which positions the game uneasily in relation to its virtual environment:

*The Long Dark* is a survival experience, and we strive for realism in many areas, but it is NOT a replacement for actual survival training or experience in the wilderness. In the end, our goal is to provide an interesting set of choices for you to play with safely. It is not a wilderness training simulation.

Regardless of how much, or in what ways it is employed, some sort of environmental orientation seems to be essential to these games, reflected in the way the players and press describe the pleasures of them. One of the early brace of features written about *Minecraft* that helped spark its initial buzz was written by John Walker of the influential website *RockPaperShotgun*. Published in September 2010, Walker specifically highlights the "freedom" and expressive dynamics of the game:

One of *Minecraft*'s most remarkable features is how different people approach it. Some see it as a giant Lego set, and set about constructing wondrous things. Others see it as a combat game, letting you create armor and weapons and fight your way through the nights. Me, I see it as an exploration and homebuilding game. (Walker, 2010)

Even within Walker's early description there is already a sense that certain players value or prioritize the survival aspects of the game, and this has become a mainstay of the genre. In his essay on *Minecraft* and its relationship to creative individuality, Josef Nguyen (2016: 489) connects the game to the tradition of 'individual survivalist' narratives like Robinson Crusoe:

Crusoe learns to view his environment as ready material at his disposal. *Minecraft* as a video game virtualizes this view of the entire world as standing-reserve… through the depiction of an entire world made of blocks to be shaped, gathered, and crafted into players' castles.

Many survival-crafting games reproduce this same narrative, with the player beginning in a particular environment with very few initial resources or capacities and little or no information about the immediate surrounding area. Over the following period, they must collect, harvest, grow, or in other ways acquire the raw materials and resources necessary to produce the tools and items that enable survival and other useful or pleasurable abilities. And while survival has always been a standing concern in many games—particularly first-person perspective games that provide an intimate connection with an avatar whose survival the player is responsible for—the form this "survival" has tended to take is a more immediate one.

In fast–paced combat games, the sense of avoiding instantaneous threats to one's survival from bullets, monsters, and other rapidly advancing dangers is primary. While these momentary threats to existence still occur in survival–crafting games (via monsters, wildlife, cliffs, etc.) the main obstacles to survival and success play out over longer intervals—dangers like hypothermia, starvation, dehydration, illness, and exposure that require long term strategies to overcome. Hunger and thirst tends to increase over time, as in games like *DayZ, ARK, Subnautica*, and the survival mode of *Minecraft*. Enemies, whether human or AI, still represent threats to the player's health, like in *ARK's* hostile dinosaurs and *Rust's* (Facepunch Studios, 2018) antagonistic players. But obtaining long-term shelter often becomes just as or even more important to securing one's survival, whether that be to escape the freezing cold temperatures as in *The Long Dark*, or to keep away from predators that come out at night in *Minecraft*. The form that shelter takes in these games varies enormously, ranging from preexisting structures players may huddle within to wait out passing threats to the ability to modify the environment itself to provide shelter, as in *Minecraft's* digging and building. This process of inscribing onto the world itself has been described as "ecocomposition" by Kyle Bohunicky (2014: 222)—the writing of "shelter" or "home" onto the natural environment, constructed from the world itself: 'rocks, trees, dirt, water and biological matter... provides a set of symbols with which they can write shelter, tools and media.' As we shall see, however, these acts are almost always embedded within crafting and resource systems that re-inscribe particular sets of relations between the player and the environment—relationships worth interrogating and critiquing.

The introduction of longer term strategic "survival" mechanics in the survival–crafting genre has shifted the dynamics and implied meanings of these games, as well as the relationships they create between players and the environments their characters inhabit. In the survival–crafting game *Subnautica* (Unknown Worlds Entertainment, 2018) the player's spaceship has crash landed on a water covered planet with you the only survivor, escaping in a pod packed with advanced technology for breaking down and synthesizing objects out of the raw materials found on the planet. Survival thus involves using the player's fabrication tools to refine simple "scrap" from the crashed ship, as well as natural resources (like copper, quartz, titanium) found in small nodes that represent different rock formations (limestone, sandstone, shale) accumulating enough of them to turn into usable raw materials to craft objects like knives and various tools. These can then be used to catch fish to cook and eat. The need to eat, and the time it takes to collect fish to cook and consume, in order to refill a hunger meter, enmeshes the player in the management of a series of dynamics of food gathering that interrupt and incentivizes certain decisions. Time taken hunting and cooking is time wasted or taken away from the accumulation of resources, exploration, and the crafting of more exciting tools and vehicles. Eating (and often drinking as well) becomes a kind of mandatory drudgery—the simulation of the necessary upkeep of maintaining a living body in a somewhat hostile environment, where one cannot simply exchange money for sustenance. The rate at which the hunger or thirst meter fills up is calibrated to the rhythms of a typical play session, and all players will find the more complex and creatively rewarding activities such as exploring and building interrupted by the need to maintain sustenance (or, as in *Subnautica*'s submarine exploration, air). Beyond instrumentalizing the body's relationship to food and water, this common survival–crafting trope frequently guides players toward stockpiling and accumulation, reintroducing economic considerations. To prevent the complete 'solving' of this problem, food often spoils, preventing the complete redundancy of these systems via accumulation. In *Subnautica*'s case, cooked fish can be 'cured' to prevent spoiling but will then dehydrate the player when consumed, solving one problem but creating another. These more processed, durable forms like salted fish or jerky reflect similar "real world" dynamics that affect all dead matter, but we would be remiss in thinking that designers employ such dynamics free from cultural predilections or other baggage. The in-game problem of spoilage tends to be solved with dynamics that mirror the imperfect solutions capitalist economics have generated for the same—like the much-maligned Soylent or ultra-processed foods. It is also worth being critical of how this dynamic plays out in the second major feature of survival–crafting games which is the crafting component of these games—also prone to the same dynamics of accumulation, and often even more reflective of economic orthodoxies.

## CRAFTING AS ACCELERATING TECHNOLOGICAL EFFICIENCY AND ACCUMULATION

Open any survival-crafting game in the genre and you will almost inevitably find a similar logic of crafting across the whole genre. Crafting involves the combination or refinement of crude or simple items or resources to form more useful, more valuable, or more complex ones. The end goal being the most efficient or expressive tools that enable the player to leave behind or greatly minimize much of the drudge work of the initial period. In some ways, this represents an adaptation of the familiar tech tree concept—present in games for many decades now and rightfully critiqued for the assumptions with which it is often inflected (Ghys, 2012; Carr, 2007). The crafting pattern established in *Minecraft* and followed by many others can be read as including similar linear assumptions about the nature of human advancement, accumulation and technology. *Minecraft* introduced the template that many others follow in this respect, in which it deploys a visual arrangement system to enable crafting of a variety of objects, with variable levels of efficiency, durability and speed associated with the different types of resources used in the construction of a tool, weapon, or armour. Arranging a T-shape out of wood creates a wooden pickaxe. Swap out the three blocks of wood across the top of the T for iron to make an iron version. While the exact crafting-via-arranging system is rarely replicated in other survival–crafting games, presenting a somewhat limited palette of options for recombination, a variety of alternatives have been imagined by others in genre. What is the same in their respective implementations is that they almost all retain a sense of tiers or levels of advancement, with increasing utility and efficiency to higher tiers of tools.

In *Minecraft*, the tools a player can craft are gated by specific design decisions that encourage this dynamic of exponential accumulation. Starting with nothing but bare hands, at first only wood or dirt may be collected (or "mined") with anything else taking an unreasonably long time. Stone mining is an excruciatingly slow process without any implements—wooden tools are the most readily accessible, however also the least durable, readily replaced by stone. Stone tools last longer, gathering faster than wooden tools, and also allow for the collection of iron for the first time. Iron, once smelted, can be turned into iron tools which collects wood, stone and iron blocks yet faster still, and is the most durable of all, save for diamond tools—a rare commodity only found in the deepest mines.

The result of this design choice in *Minecraft* is a clear tool and resource hierarchy beginning with one's bare hands and ending with diamonds. The limited durability and slow speed of mining that occurs at the lower rungs of the tool hierarchy restrict player agency and expression, hampering the speed of resource gathering and deployment, and giving play sessions a very loose and optional structure through incentivizing efficiency. The decisions about how these hierarchy of tools appear is obviously informed by real world characteristics and considerations: diamond is one of the hardest natural substances in the world, is used in commercial mining applications, and forms a useful mental association for the player to understand this technology as the pinnacle. It immediately makes sense for it to appear as the ideal or best tool, conforming to players' expectations based on knowledge of natural behaviours. *Minecraft* may have set the template in this regard, but others in the genre have refined and expanded it.

In *Subnautica*, for instance, crafting is performed by selecting from a menu of blueprints rather than a visual grid. *Subnautica* begins initially with access restricted to crafting simple wetsuit apparatus including flippers, proceeds with enough accumulation of resources to crafting individual propulsion systems, and finally fully enclosed submersibles that can act as mobile factories. By this point the scope of player agency and the actions afforded them has expanded enormously. Each increase in technological complexity increases the speed of the player's passage through the world as well as the rate of resource accumulation. This is important as each new unlocked fabrication plan often requires exponentially more resources, or at least previously unobtainable ones. In this way, *Subnautica* mirrors the intensification and acceleration of resource use since industrialization that is one of the principal features of modern capitalism. For *Subnautica*, the player's technological journey from simple tools to complex machinery cannot avoid comparisons to increasing resource and carbon intensity as economies develop and standards of living increase. Because there is only one person consuming resources in *Subnautica*, and because the player's long-term goal is to essentially leave the planet behind rather than to make a life for themselves on it, the tragedy of the player's resource intensity largely recedes into the background. It would be rare for the resource consumption of the player to become a 'problem' for the ecological carrying capacity of the planet in *Subnautica* as certain resources respawn, and there are always more areas to plunder.

The game then seems to simulate (in a sort of hyper-accelerated caricature) the unsustainability of technological development, one of the major contributors to the problems facing our planet today. *Subnautica* is by no means alone in producing this dynamic, however. On the contrary, it is a central feature of the genre, one that undoubtedly provides a great deal of enjoyment for players (myself included) to engage with. There are real pleasures to accumulation, the satisfaction of having overcome scarcity, achieving comfort and security that developers have rightly identified and reproduced in many games like it. In this way, I don't wish to single out the game for critique as criticism of the developer, but rather to illustrate the common shared features across almost all the games in the genre and the questions that we must ask about the kinds of pleasures and desires these games are tapping into.

In *ARK*, a slightly different pattern is used again, one that involves gaining experience and levelling up to unlock blueprints for more complicated tools—but the same dynamic of accelerating accumulation is in full force. In *The Forest (Endnight Games 2018)*, the player peruses a menu of craftable items (visualized as a survival handbook with pages and illustrations) and that feature the same amplification of agency and increasing efficiency as these tools improve—a sled that can be loaded and dragged with several times the amount of wood that an individual player could carry previously, for instance.

In most of these games then, the crafting progression with its hierarchy of materials and tools comes to reflect a simplified version of the ladder of technological progress. One of its main purposes is to provide something of a goal or structure in which the freeform play of building, survival, and self-expression can unfold in an otherwise open or undirected world. This same dynamic also allows for the rewarding of players for their time spent in the game through increasing the rates of return on resources. A negative review on the Steam page *ARK* makes this point clear. Steam user "Morticielle," who at the time of posting in January 2016 had recorded over 3400 hours of playtime in the game, claimed that their "love" for the game had been "dying" for several months due to changes to high–end, difficult to acquire dinosaurs introduced in a previous patch. Specifically, the user claims that:

> Animals which were used to farm metal, stone, wood etc. now have a much much lower yield than before. This has as a consequence that getting resources for crafting now becomes a nuisance. [sic]

Here, *ARK* player Morticielle uses explicitly economic language to describe feelings of frustration over a decreased ratio of time vs reward. In other words, since this latest patch they have been taking a haircut on their metal, wood and stone rate of return and are threatening to take their time investment elsewhere. Many players understandably gravitate towards the most efficient tools in order to spend more time on enjoyable and rewarding tasks that enable their self-expression, creativity, and control over in-game environments. The less time spent on the work necessary for enabling such self–expression the more time there is for more enjoyable activities—just like when you aren't being interrupted by the need to find food every so often because of sufficient accumulation. The last several years has seen a body of work taking the concerns of players about the kind of "grind" required, considering what constitutes work in games, occasionally referring to it as "playbor" (Kücklich, 2005; Dyer–Witheford & de Peuter, 2009). In Morticielle critical review on Steam, they express an implicit preference for spending more time in *ARK* planning and constructing and less time labouring away on resource gathering. There is still clearly room for more detailed understanding of struggles over control of the means of production (so to speak) in digital games.

In summary then, the crafting dynamics of survival-crafting style games hew more closely to and prompt from users more thought and reflection about economic ideas and experiences than environmental ones. Despite the environmental correspondence suggested by terms like realism and emergence, and the replication of bodily dynamics like hunger and thirst, they almost invariably end up replicating capitalist notions of return on investment and increasing rates of accumulation (if in rather simplified, cartoonish ways). Under capitalism it is often said that it "takes money to make money," and in survival–crafting we might alter that formula slightly: it takes time to save time. These dynamics of accumulation and reward structure the rhythms of play in these games, providing what many players find an enjoyable and rewarding pattern of effort and reward that is a classic feature of game design, reinstated here in a regime that centers itself on time and resources rather than, say, the traditional rewards of loot and XP.

These are clearly constructions worthy of closer analysis, particularly as the dynamics created though a hierarchy of tools and resources reflects a form of simplistic technological determinism. Certainly, there are few opportunities for moments of revolutionary discovery in a survival–crafting game, very little opportunity for the reconfigurations and reorderings of social worlds that are often instigated by real technological development (both positive and negative). Furthermore, we can compare the trajectory of increasing degrees of player "mastery" over the landscape over time with a similar trajectory embedded in Western and colonial thought, with a deep (if simplified) connection to metanarratives about the triumph of humanity over untamed Nature. By typically presenting verdant, unspoiled landscapes for the player to reshape and repurpose as they see fit, the genre reproduces what feminist ecologist Val Plumwood (2003) calls the "mastery of nature" narrative. As enjoyable as they are, survival-crafting games typically simulate (again, in cartoonish form) what Plumwood (2003: 2) calls the same 'problematic features of the west's treatment of nature which underlie the environmental crisis, especially the western construction of human identity as 'outside' nature.'

So how do we reconcile this situation? Despite the claims we saw earlier about *Minecraft*'s correspondence with real environmental and ecological credentials (many of the features which appear in other games in the genre), despite the fact that we might be tempted to laud this genre of games as an encouraging development, given the pleasant inclusion of bucolic settings within many of these games, how do we square that they also involve taming landscapes and extracting resources, and reproduce many of the same ideas that environmentalists and ecologists have been struggling to dislodge from the popular consciousness since the modern environmentalist movement began in the 1960s? This conundrum has also been highlighted by Chang, (2011: 60) who suggested long before the current flourishing of the survival–crafting genre, that 'game designers have yet to develop more sophisticated rules for interaction between play-ers and game environments.' As a result, 'games naively reproduce a whole range of instrumental relations that we must reimagine' (Chang, 2011: 60) evidence of which there is still plenty of. Examination of the survival–crafting genre seems to suggest that hopes for a greater sophistication in player–environmental interactions have yet to be fulfilled in a particularly meaningful way, at least in this genre.

## THE ECOLOGICAL GAME

What hope is there, then, for the realization of a truly ecological game? When it seems that even the survival–crafting genre, lauded for its repro-duction of natural dynamics and environmental interactions, and surely one of the most ecosystemic and 'emergent' genres of game—is itself thoroughly suffused with the same thought and dynamics that underpin the current ecological crisis? What is, or would, a truly ecological or eco-critical game resemble? Recall the ecological thought—which recognizes that "everything is connected," and not just in a vague new-age way, but actually connected via links and interrelationships that reflect both mate-rial and conceptual connection. As Morton (2010) reminds us, there is no "away" or "outside" for our waste, our carbon emissions, no place for them to be sent to simply disappear. Thinking ecologically 'isn't *like* think-ing about where your toilet waste goes. It *is* thinking about where your toilet waste goes' (Morton, 2010: 9).

If our games are to be ecological then surely we must also think ecologically about them, and which opens up a host of new and unforeseen questions. Firstly, it seems to me, that thinking ecologically means making visible all the waste involved in a game—from the carbon dioxide emissions from development, to the materials used in getting the game to players all around the world (plastics, bunker and aviation fuels, and so on) to the materials in the devices required to play the game. Once we begin to think about these emissions and their impacts, and start to think in a connected way about the consequences of a game for the planet, we see immediately that games are distinctly intertwined with the rest of the climate crisis, responsible in some small way for increasing $CO^2$ levels in our atmosphere, and that this process is *not* separate or distinct from what happens in a particular game. From the lush biomes of *Minecraft*, to the beautiful underwater vistas of *Subnautica*, these fantastic visions of environments come with an often-invisible cost to non-virtual environments. Once we acknowledge this it becomes incredibly difficult to avoid seeing other connections as well, and to follow the possibilities presented by this kind of ecological thought about games. By tracing the flows of energy, emissions, plastics, metals and other materials in, around and through games themselves we find out so much that was previously obscured when our focus is solely what happens on-screen. Issues that might have previously been considered utterly disconnected from the pressing of a button on a controller, typically escaping attention in our analyses and close readings of games, are now front and center—unavoidably linked through chains of interactions that are material, logistical and above all ecological. From plastics, to energy, to carbon emissions, workstations, server rooms, transport and logistics of people and things all around the world in this increasingly interconnected and global industry—these are all part and parcel of the ecological implications of making particular games in this hugely impactful industry.

For much of its history, the field of game studies was largely content to keep research questions and analysis confined to the game experience—fair enough, as far as it goes, for the early stages of a new field. But this focus remained stubbornly fixed, and has all too rarely expanded to encompass the full breadth of the industry, its sites of production and play, meaning that game studies (and perhaps the games industry itself) still remains largely unaware of the full nature and extent of material and environmental entanglements that extend far beyond even the situated context of play. The broad strokes perhaps are known—with problems like e-waste, energy intensity, and so on appearing on the radar—but the details, the specific damage, the pain and suffering that industrial game production causes for both people and the planet, as well as what is to be done about it, remains underexplored. Close readings or analyses of mechanics, story or narrative, genre and convention, despite an encouraging trend to include wider cultural and political consequences of games, still do not often (or easily) connect to matters of material and ecological significance. It is not at all impossible that future analyses (even reviews) of games include evaluations of their global harms and carbon impact. The glimmer of this sort of concern appears around, for instance, questions of crunch and overtime—the *human* cost involved in producing award-winning game experiences like *Red Dead Redemption 2* (Legault & Weststar, 2017; Cote & Harris, 2021). But all too often we stop at the human dimensions. This largely reflects the same view that treats digital games as just another part of the distinct and siloed realm of culture. Games are not just culture, they are part of the world. We should continue the same thought, and ask ourselves what are the *nonhuman* impacts of a game as well? We could ask ourselves "what would a nourishing terrain look like" in a game, following Rose's (1996) articulation of Australian Aboriginal notions of care for country and environment? At the very least, we can say what it *would not* look like—a game that embodied this sense of care for the land itself would not participate in the further degradation of country, and the using-up of natural resources—whether hydrological, atmospheric, or mineralogical.

It has been years since Apperley and Jayemane (2012) noted the rise of a burgeoning 'material turn' in games studies, which brought much overdue focus on issues like production and labour issues in the game industry, as well as a greater focus on the code and hardware underpinnings of modern games. There is now a small but growing body of work, both popular and scholarly, analyzing games from materialist and environmentalist perspectives, from games as consumer objects and commodities, (DyerWitheford & de Peuter, 2009) and games as the intimate coupling of technology and human bodies, (Ash, 2013; Keogh, 2018) to games as gendered and racialized spaces (Chess, 2017; Gray, 2014, 2020). Gaming happens in private homes behind closed doors, and in public on private screens, but is generally more accessible to researchers. Game development happens in private settings which researchers often have difficulty accessing, explaining perhaps partly why the industrial aspects of game production have been less frequently studied (Bulut, 2020; O'Donnell, 2014). Attending to the particularities of energy demands of particular games, in the particular locations they are played, as well as attending to the demands of the individual devices they are played on, points towards one way of establishing some of these missing connections.

In his essay "A critique of play," Sean Cubitt quotes geneticist R.C. Lewontin for an explanation of the relationship between organism and environment, suggesting its relevance for emergent game forms specifically: 'because organisms create their own environments we cannot characterize the environment except in the presence of the organism that it surrounds' (Cubitt, 2009). The conclusion Cubitt (2009) draws is that 'an ecological game is then one in which the act of externalizing and objectifying the environment as other is broken down by insisting on the mutuality of production, the interaction of multiple users to produce an evolving ruleset.' Does this not describe the kind of game that refuses to ignore the emissions made through its development, through it being played, and not just what happens within the confines of the games inner code?

This perspective challenges the prevailing approach of much game analysis which treats the boundaries of what happens within the game itself, or perhaps the boundaries of what happens between game-and-player, as the appropriate (sometimes even the *only*) unit of analysis. Ecological thought, however, challenges the very validity and coherence of these boundaries, entreating us paradoxically to think both bigger and smaller at the same time. I have argued elsewhere (Abraham & Jayemanne, 2017) for the useful application of a suite of lenses in the analysis of ecological representation in games in the moment of gameplay. My co-author and I concluded by highlighting the partial and fragmentary nature of any such analysis, emphasizing the way games escape the boxes we place them in: 'none of these four modes really capture the potential of how the weird assemblages we call videogames can deal with the weird event we call climate change' (Abraham & Jayemanne, 2017: 15).

On a moment-to-moment level, games like those in the survival crafting genre may succeed or fail at representing or simulating an ecological or even ecocritical event or narrative, but at the same time, the game object or experience exceeds these boundaries, if we care to look. Though uncommon, there are games that draw our attention to this, like the *Metal Gear Solid* (MGS) series which often gestures beyond the bounds of the game experience with fourth–wall breaking moments. These serve to remind that the game knows it's being played, that the player is in some particular location, often highlighting the specific console or device itself. In the original *Metal Gear Solid* title for PlayStation, during the much-loved Psycho-Mantis fight, I still recall having no clue what to do, and being genuinely fooled into believing that my TV had changed channel— the black screen with green lettering "HIDEO" close enough to my exact Sony television's "VIDEO" font and appearance; a deeply confusing and confounding experience. This movement towards the player/hardware interface presents itself as an example of, and opportunity for, ecological thought even though this might not seem to conform to our green sensi-bilities. One can imagine similar movements and attention directing fea-tures: logging and displaying the power consumption involved in playing for a given game session; reminders that the development of the game was done in a carbon neutral way; features that make the game dynamically responsive to local emissions intensity, and so on. These are extremely rudimentary suggestions, but we shall discuss more in Chap. 6. In some cases these might be small gestures, to be sure, but they and others like them are a small step towards a different kind of engagement with games.

We could, in fact, go even further than this and consider whether there might be something about the nature of interaction itself which seems in accord with ecological thinking. The concept of interaction harbors a deep ambivalence or conceptual ambiguity about what it refers to exactly, reflecting a confusion in distinctions between the material and the cultural in games. Salen and Zimmerman's (2004) discussion of interaction in *Rules of Play* is unable to reconcile the disparate ways in which the term is conceptually deployed to present a unified definition. The closest they come to a generalized definition of interaction is when they say that 'inter-activity simply describes an active relationship between two things' (Salen & Zimmerman, 2004: 58). This almost unhelpfully broad definition has real and obvious resonances with Morton's (2010) ecological thought as well. But if tracing "interaction" is at least like thinking ecologically, then an analysis of mechanical interactions and the meanings they create, as well as the ideologies and attitudes about nature that they embody remains a valid avenue of critique. If we look at the ideas and ideologies embodied in interactions within games—in other words, the nature of relationships between player and world, player and objects, player and other things as foregrounded or backgrounded by design—then we are just as much doing ecological thinking. The trick is not to stop there, but to keep on pushing further, by tracing yet more ecological connections.

Is it possible then to consider even first–person shooter (FPS) games to be in some sense "ecological"? Is it not a deeply ecological activity to influence the world (even in a destructive, negative sense) when the player reaches out and catalyzes interactions between bullet and world, when bullet hole decals appear on a texture, and when objects, players, systems respond in kind? Would a more 'ecological' FPS include a more durable record of the (inter)actions of players in game spaces, or present a record of the energy used by a player in the operation of data centers? Surely it is both. Chang (2020: 71) notes the same correspondence between interaction and ecology, noting that 'the simple idea that everything around us may somehow be impacted by our presence, let alone our actions, is a tenet of ecological thinking.' Chang (2020) however is concerned by the one-dimensional nature of said interactions in many popular games, and is disappointed,

> when a game's celebrated interactivity amounts to little more than blowing up shooting, torching, or otherwise causing gross and systemic destruction. What if, instead, games like these underscored the ways in which damaging acts inevitably recoil on their perpetrators, as well as less spectacular and more insidious effects, ranging from contamination to deterioration?

To me, however, I see less importance in emphasizing distinctions between "good" and "bad" ecological interactions in game worlds. Game interactions are simply too weird, and the affordances (and implied ontologies) of game worlds will never approach the level of ecological complexity of the real world. I think it might even be important to affirm that there is hardly anything more or less artificial about a virtual bullet versus, for instance, a virtual dandelion, a virtual rabbit, or even a virtual human being—this is thinking ecologically! From this perspective, a virtual explosion on a screen powered by carbon-free energy, could be more ecologically-oriented, and be doing *materially* less harm than a similar climate friendly story in a game that it is played on a machine powered by fossil fuel generation.

Perhaps the challenge then is to find ways to connect ecological thinking inside the so-called "magic circle" of the game space with the wider ecological connections that games entail via their production and consumption. (Möring, 2019) A social media post that went viral in gaming communities online in late 2019 managed to do just that, stating wryly that 'lamps in videogames use real electricity.' This is both true and a pithy reminder of the challenge facing the ecologically-minded game maker. It's not always easy, or possible, to know where the real electricity for your virtual lamp is going to come from. Games and game design need to somehow begin to account for all of this, despite it being perhaps a bizarre thought to most developers at present. The waste and harms generated by gaming being 'elsewhere', conveniently invisible at the point of consumption and play, does not make them any less, just less visible, more pernicious and harder to track.

In the conclusion to Chap. 2 I argued that instead of trying to persuade or alter the hearts and minds of the climate sceptic player it is both easier and more essential to change the very world around them. This is precisely the benefit we achieve when we succeed in thinking about and acting ecologically. The ecological game is one that attends to its own material conditions, its development and production, its existence and even its afterlife, and takes an active role in mitigating the harms produced. By doing this the climate-skeptic is side-lined, no longer needing their participation to achieve climate goals. I do not mean that climate skeptics would be unable to play an ecological game, but rather that the climate skeptics *no longer pose a problem to the ecological game.* The truly ecological game is already aware of its carbon emissions, has already reduced its carbon footprint, has already decoupled itself from the resource extractive aspects of the games industry. From the vantage point of the developers of such a game, it no longer matters much whether or not a game can or does change a climate skeptic's mind. What a joy to be freed of the burden of what might otherwise be an impossible task, having instead fixed ourselves on the very real possibility of changing the world instead. Suffice to say, the truly ecological game is a challenging vision, and the first (hopefully of many) might still be some way off. In this way the ecological game is both an ideal, a target to aim for, and hopefully one day an *actual* game that embodies these ideals. As we shall see in Chap. 6, there are already some developers approaching this goal—the first true ecological game may not even be far off. It may also look very different to what we expect. For all else we can say about it, however, it is now clear that the truly ecological game must be, first and foremost, a carbon neutral one.

## CONCLUSION

This chapter began with a discussion of the ecological resonances present in game discourse terms borrowed from or connected to ecological and environmental ideas—from environmental storytelling to the concept of emergence, to the way games like *Minecraft* and others in the survival-crafting genre seem to embody environmental dynamics. The chapter ended with a return to the concept of interaction itself, which has key resonances with ecological thought but which remains difficult to reconcile in its typical application. Connecting our everyday notions of interaction with the kind of truly open-ended ecological thought, and the impossibility of any true end to ecological connections, remains a challenge. Even if we focus our analysis on the dynamics of the games that position themselves as more 'world' focused through emergence and similar terms, we still struggle to find them any more ecological than another, except perhaps in a thematic sense. Or, perhaps more accurately, games like those in the survival–crafting genre are ecological only to the same extent that *any game, any interaction* is ecological. As this chapter has argued, much of the focus of the survival–crafting genre of games, and the systems and dynamics they enroll players into, boils down to economic activity and logics of accumulation, arguably doing little to prompt player thought or action in an ecological direction. For these reasons, the genre and the discourses that position it as better on environmental grounds deserves slightly more skepticism than it usually receives. But the same applies to all games, all deserving greater interrogation of their relationships between representational elements of game environment and the real material stuff implicated by their production.

With a few exceptions, analogue games like Monopoly, Scrabble, Carcassonne, Settlers of Catan, and so on can all be played without electric power provided one is furnished with a well-lit room, a bit of free time, and access to the necessary components. Likewise, physical and sports games, at least in their ad-hoc, amateur forms can be played without electricity, and often are. Digital games, however, do not have that luxury. Digital games *cannot* escape their tether to electrical energy, and some sort of computational hardware to run on. Every watt of power used by gaming devices is one watt of power still more likely than not generated by burning coal, gas and other fossil fuels—though this is changing. Every game that gets made is made in a workplaces that likewise draws power from those same fossil fuel powered energy systems. Every disc that gets transported to a shop for sale is made somewhere else, trucked, shipped, or flown to that shop, using up plastics and fossil fuels in the process. Every console that gets switched on has been made out of intricate assemblages of precious materials like gold, indium, copper, palladium, which have to be dug out of the earth's crust and refined to a shocking degree of purity through processes of incredible energy intensity. At virtually every stage we find emissions of carbon dioxide and other greenhouse gases that are trapping heat in the atmosphere, accelerating climate change and driving more and more extreme weather. Our first and final goal then *must be* achieving a carbon neutral games industry. From top to bottom, from the earliest stages of pre-production, from planning the next game and the next game consoles, from the distribution center to the data center, to the very last light of a device once it is superseded through obsolescence—at all points along the lifecycle of a game and gaming devices we must identify and neutralize greenhouse gas emissions as much as possible. A truly ecological game can only be one that is aware of its existence as part of the material world, the harms it causes to a range of humans, plants, animals and the planet, and takes active steps to eliminate them.

There are some encouraging signs from individuals and companies beginning to do this, and we shall see more in the chapters that follow. Nguyen (2016) has already argued two games, *Little Inferno* by Tomorrow Corporation and *Phone Story* by Molleindustria, succeed in this process of awareness of their own materiality, of both the game and its player's active participation in harmful processes. But games must go beyond awareness. A few are beginning to. In August 2020, the gaming merchandise manufacturer Iam8bit partnered with Melbourne based developers House House, makers of *Untitled Goose Game*, to produce the 'Untitled Goose Game Lovely Edition'—the first and only physical edition of the game, expressly designed to use sustainable packaging. In a twitter thread, the developers explained their decision to avoid plastics, instead use an all-recycled cardboard material, printing that contained 'no harmful inks', and a biofilm to replace plastic shrink-wrapping. Inside the box, stickers were made from sugar cane waste and a map of the town was printed on Forest Stewardship Council-certified paper. The disc itself, presumably, remains polycarbonate with no mention of a replacement for this essential plastic component with sustainable materials, however even so it remains a positive step towards the truly ecological game. In the next chapter we will meet game developers who are already working towards making changes to their energy use through efficiency, carbon offsets, renewable power, and other initiatives.

The remaining chapters of this book outline the urgent task of understanding what the ecological game actually means for both the planet and the games industry, documenting the scale and scope of emissions and other environmental harms that all games are going to have to engage with. To address the climate crisis, all games will need to become ecological games. This will be a job that cannot be done solely in theory or books like this one, but will need to be taken up practically—by game makers, scholars, journalists and activists as well as game players themselves. This needs to become a central pillar of the modern games industry, and it needs to happen now. The chapters that follow offer an attempt at filling in the picture of where and in what sorts of amounts these carbon emissions are occurring, what is involved both in terms of the scale and nature of the challenge, and offer some suggestions for how they might be reduced or eliminated altogether.

**Acknowledgments** Parts of Chap. 3 have appeared in *Trace: A Journal of Writing, Media and Ecology*.

## References

Abraham, B., & Jayemanne, D. (2017). Where are all the CliFi games? Locating digital games response to climate change. *Transformations, 30*, 74–94.

Adams, E. (2009). *Fundamentals of game design*. Pearson Education.

Adams, J. (2012). Inquiry learning? Try a game. Teaching boys at the coal face: Mining key pedagogical approaches. *Action Research Report 2012*. International Boys' Schools Coalition. (pp. 5–24).

Alexander, L. (2013, November 16). Spector: Go emergent – Game design is not all about you. *Gamasutra*. https://www.gamasutra.com/view/news/204942/Spector_Go_emergent__game_design_is_not_all_about_you.php. Accessed 30 Mar 2021.

Apperley, T., & Jayemanne, D. (2012). Game studies' material turn. *Westminster Papers in Communication and Culture, 9*(1), 5–25.

Ash, J. (2013). Technologies of captivation: Videogames and the attunement of affect. *Body & Society, 19*(1), 27–51.

Ballantyne, A. G. (2016). Climate change communication: what can we learn from communication theory?. *Wiley Interdisciplinary Reviews: Climate Change, 7*(3), 329–344.

Bohunicky, K. M. (2014). Ecocomposition: Writing ecologies in digital games. *Green Letters, 18*(3), 221–235.

Bulut, E. (2020). *A precarious game: The illusion of dream jobs in the video game industry.* Cornell University Press.

Carr, D. (2007). The trouble with civilization. In B. Atkins & T. Krzywinska (Eds.), *Videogame, player, text.* Manchester University Press.

Chang, A. Y. (2011). Games as environmental texts. *Qui Parle, 19*(2), 56–84.

Chang, A. Y. (2019). *Playing nature: Ecology in video games* (Vol. 58). University of Minnesota Press.

Chang, A. Y. (2020). Rambunctious games: A manifesto for environmental game design. *Art Journal, 79*(2), 68–75.

Chess, S. (2017). *Ready player two: Women gamers and designed identity.* University of Minnesota Press.

Cote, A. C., & Harris, B. C. (2021). The cruel optimism of "good crunch": How game industry discourses perpetuate unsustainable labour practices. *New Media & Society.* https://doi.org/10.1177/14614448211014213

Crisp, S. (2015). *The realist videogame as performative text.* Honors Thesis, RMIT University.

Cubitt, S. (2009). A critique of play. *Refractory: A Journal of Entertainment Media,* 16.

Duncan, S. C. (2011). Minecraft, beyond construction and survival. *Well Played: A Journal on Video Games, Value and Meaning, 1*(1), 1–22.

DyerWitheford, N., & De Peuter, G. (2009). *Games of Empire: Global capitalism and video games.* University of Minnesota Press.

Ekaputra, G., Lim, C., & Eng, K. I. (2013). *Minecraft: A game as an education and scientific learning tool* (pp. 237–242). ISICO.

Galloway, A. R. (2004). Social realism in gaming. *Game Studies, 4*(1) http://gamestudies.org/0401/galloway/

Ghys, T. (2012). Technology trees: Freedom and determinism in historical strategy games. *Game Studies: The International Journal of Computer Game Studies, 12*(1).

Gray, K. (2014). *Race, gender, and deviance in Xbox live: Theoretical perspectives from the virtual margins.* Routledge.

Gray, K. L. (2020). *Intersectional tech: Black users in digital gaming.* LSU Press.

Hobbs, L., Stevens, C., Hartley, J., Ashby, M., Lea, I., Bowden, L., Bibby, J., Jackson, B., McLaughlin, R., & Burke, T. (2019). Using Minecraft to engage children with science at public events. *Research for All, 3*(2), 142–160.

James, W. (1909). *A pluralistic universe: Hibbert lectures at Manchester college on the present situation in philosophy.* Longmans, Green and Co.

Jenkins, H. (2004). "Game design as narrative architecture", in Wardrip-Fruin, N. and Harrigan, P. *First Person: New Media as Story, Performance, and Game.* Electronic Book Review.

Keogh, B. (2018). *A play of bodies: How we perceive videogames.* MIT Press.

Kücklich, J. (2005). Precarious playbour: Modders and the digital games industry. *The Fibreculture Journal, 5*(1).

Lane, H. C., & Yi, S. (2017). Playing with virtual blocks: Minecraft as a learning environment for practice and research. In F. C. Blumberg & P. J. Brooks (Eds.), *Cognitive development in digital contexts* (pp. 145–166). Academic Press.

Legault, M.-J., & Weststar, J. (2017). Videogame developers among 'extreme workers': Are death marches over? *E-Journal of International and Comparative Labour Studies, 6*(3), 1.

Möring, S. (2019, August 9). Expressions of care and concern in eco-critical computer game play. In *Digital games research association annual conference 2019*. Ritsumeikan University.

Morton, T. (2010). *The ecological thought*. Harvard University Press.

Nguyen, J. (2016). Minecraft and the building blocks of creative individuality. *Configurations, 24*(4), 471–500.

O'Donnell, C. (2014). *Developer's dilemma: The secret world of videogame creators*. MIT Press.

Philips, A. (2014). Queer algorithmic ecology: The great opening up of nature to all mobs. In N. Garrelts (Ed.), *Understanding Minecraft: Essays on play, community and possibilities*. Jefferson.

Plumwood, V. (2003). *Feminism and the mastery of nature*. Routledge.

Rose, D. B., & Australian Heritage Commission. (1996). *Nourishing terrains: Australian Aboriginal views of landscape and wilderness*. Canberra.

Salen, K., & Zimmerman, E. (2004). *Rules of play: Game design fundamentals*. MIT Press.

Sharp, L. (2017). The geology of minecraft. *Teaching Science, 63*(1), 14–17.

Shinkle, E. (2020). Of particle systems and picturesque ontologies: Landscape, nature, and realism in video games. *Art Journal, 79*(2), 59–67.

Skrebels, J. (2016, March 18). ARK: Survival evolved's poop button has a dark origin story. *IGN.com*. http://au.ign.com/articles/2016/03/18/arksurvivalevolvedspoopbuttonhasadarkoriginstory

Spokes, M. (2018). 'War…war never changes:' Exploring explicit and implicit encounters with death in a postapocalyptic gameworld. *Mortality, 23*(2), 135–150.

Walker, J. (2010, September 27). Cave time: Feeling safe in Minecraft. *RockPaperShotgun*. https://www.rockpapershotgun.com/cave-time-feeling-safe-in-minecraft. Accessed 2 Apr 2020.

GAMES

Bethesda Game Studios. (2011). *The elder scrolls V: Skyrim*. Bethesda Softworks.

Crytek, Ubisoft Montreal, Ubisoft Toronto. (2004–2021). *Far cry series*. Ubisoft.

Endnight Games. (2018). *The forest*. Endnight Games.

Facepunch Studios. (2018). *Rust*. Facepunch Studios.

Hinterland Studio. (2017). *The long dark*. Hinterland Studio.

Ion Storm. (2000). *Deus Ex*. Eidos Interactive.

Mojang. (2011). *Minecraft*. Mojang, Microsoft Studios, Sony Interactive Entertainment.

Re-Logic. (2011). *Terraria*. 505 Games.

Studio Wildcard. (2017). *ARK: Survival evolved*. Studio Wildcard.

Unknown Worlds Entertainment. (2018). *Subnautica*. Unknown Worlds Entertainment.

# CHAPTER 4

# How Much Energy Does it Take to Make a Videogame

Exactly how much energy does it takes to make a modern video game? The answer is likely as complicated as the games themselves, and a precise number for any given game is likely near impossible, perhaps achievable only with the full commitment of a deeply committed and prepared game development studio, publisher, and even players, and would involve a herculean effort applying best practice emissions accounting from the very outset of development. Even then, exact figures may prove elusive, being something of a moving target, especially for the increasing number of games-as-a-service titles receiving ongoing updates and support. In any case perfectly exact numbers are not essential in the grand scheme of things. Digital games are, as we shall see, a fraction of a percent of total global emissions (though that should not give them carte blanche) and it seems doubtful to me that the embodied emissions of a particular game are ever going to become the deciding factor in whether a player picks up a game or not. That's simply now how gamers pick and choose the games that they play. (Westwood & Griffiths, 2010; Erol & Çirak, 2020) What is possible, however, is that the social conscience of developers is expanded to encompass emissions, much in the way that the mainstream games industry has come to accept the importance of diversity. Much like how this change has been driven by a decade or more of activism from women, people of colour, and LGBTI people in and around the games industry (Golding & van Deventer, 2016; Ruberg, 2020; Phillips, 2020; Shaw,

B. J. Abraham, *Digital Games After Climate Change*, Palgrave Studies in Media and Environmental Communication, https://doi.org/10.1007/978-3-030-91705-0_4

2015) I would hope the next decade sees a similar climate consciousness emerge among game developers. Anecdotally, I see the beginnings of this already, with evidence of this throughout the chapter.

So how much energy is used in the workplaces where games are made? How much is used across the entire sector and how readily can this energy by decoupled from fossil fuels? While deceptively simple questions, the answers are clouded by many factors. For example, at the firm or studio level these include: how many employees are present, at what kinds of workstations, running at what sort of utilization rate, for how long, at what times of the day, supported by what sort of networking, lighting, heating and cooling systems? In a truly ecological fashion, these questions sprawl and touch on others—how long does the air conditioning or heating system stay on, and is this tending to increase as the world warms? Do these hours of operation increase during periods of 'crunch' and overwork? How often are workstations upgraded or replaced, what sorts of emissions are embedded in these machines, and what happens to them once they are discarded? Are there on-site servers and server rooms, and how are they cooled? Are motion capture performers, voice over artists and other actors flown out to particular studio spaces? How are those recording spaces powered, lit, heated, cooled? How big of an orchestra is used to record a game's music, how long are the recording sessions, how much power does all the equipment like mixing desks, microphones, and amplifiers use? How many flights does a given particular development team end up taking over the length of a development cycle? How often do developers travel to conferences and conventions to network, sharpen their technical skills, and learn from each other? What sort of distances are involved in travelling to different parts of the world to market and present new games like at E3 and other trade shows? How much paper and plastic marketing material, game manuals, cloth maps and other sundries are printed and distributed alongside the plastic boxes with discs in them, sent to all the corners of the globe? The answers to all these questions have some bearing on how easy or difficult it will be to achieve a carbon neutral games industry, what sorts of actions are required, and what to prioritize. Eventually, and the sooner the better, all these emissions will need accounting for.

The first step then is to start filling in the picture of the overall energy intensity of game development and make some best guesses at where the most effective actions are to be taken. The first section of this chapter presents the results of a small study of game developers who have self-reported their energy use, and includes data from studios of a variety of sizes. The goal is to be able to offer a baseline for comparison, and a fairly simple way to extrapolate to an approximate overall carbon footprint of game development activity. What these figures provide is a benchmark for studios to self-assess against, by measuring whether they are over or under the figures of their peers and industry leaders. The primary metric here is a figure of kWh's of electrical energy used per annum, per employee—a measure of power used over time. From that, estimates of emissions intensity for the amount of power used can be made, based on the emissions factor of a particular region's power generation. It is hoped that with these baselines in mind, studios both large and small can measure themselves against leaders in the industry simply by using the measurements provided on their electricity bill. By taking something as simple as an electricity bill it is possible to obtain a fair estimate of a company's direct carbon emissions, and later in the chapter I use these estimates and the per-employee emissions figure to extrapolate a fairly rough estimate of the emission intensity of the entire games industry. By looking at how much electrical energy is used at a given studio, and by making some reasonable assumptions based on similar studios, we can extrapolate a rough sense of how much is being emitted across the sector.

The one advantage we have in the face of the complexity of this task is (or was, until the 2020 pandemic rapidly necessitated widespread work from home practices) the relatively clear boundaries of the game studio which provides a convenient scope for emissions accounting. Thomas et al. (2000, 19) notes in the UN's greenhouse gas emissions indicator guidelines, however, that 'defining a company's boundary for the purposes of accounting for GHG emissions is notoriously difficult.' Corporate structures can be opaque, with the largest multinationals existing across multiple offices, cities, regions and counties. The World Resources Institute (WRI) and the World Business Council for Sustainable Development (WBCSD) have jointly developed the Greenhouse Gas Protocol (WBCSD, WRI, 2004) building on the work of Thomas et al. (2000) becoming the world's most widely used GHG accounting standard. The Greenhouse Gas Protocol suggests counting two main types of emission: Scope 1, Scope 2 and an optional category of Scope 3 emissions. Scope 1 emissions are *direct* emissions the organization makes by burning fuels, like operating boilers, furnaces, vehicles, and so on. Most games development companies are unlikely to make much use of Scope 1 emissions themselves—with the exception of some located in cold climates. Scope 2 emissions are where we would expect most game developers to find the bulk of their emissions, with the protocol defining Scope 2 as follows:

> Scope 2 accounts for GHG emissions from the generation of purchased electricity consumed by the company. Purchased electricity is defined as electricity that is purchased or otherwise brought into the organizational boundary of the company. Scope 2 emissions physically occur at the facility where electricity is generated. (WBCSD, WRI, 2004: 25)

Any emissions from electrical energy bought from a power utility and used by a game developer is considered Scope 2 emissions, and should be attributable to the company itself despite physically occurring elsewhere, like in a coal fired power plant, a gas turbine, etc. Lastly, Scope 3 emissions are 'an optional reporting category' for emissions that occur as 'a consequence of the activities of the company, but occur from sources not owned or controlled by the company' (WBCSD, WRI, 2004: 25). The use of products made and sold by the company (i.e. the games themselves and the emissions from electrical energy used to play them) is considered Scope 3 emissions—however few companies report these figures (at least for now) and ultimately tend to leave Scope 3 emissions out of the picture as the end user's responsibility. There is a larger question here about the legitimacy of these accounting boundaries and demarcations of responsibility, however they are mostly beyond the scope of consideration here. There are however some developers beginning to take responsibility for Scope 3 emissions— the end users and players of their games. London based Space Ape Games, as well as Finnish company Rovio the developers of the *Angry Birds* mobile games, have attempted to calculate and provide carbon offsets on behalf of the emissions of their players—but these are exceptions to the norm, and I discuss the feasibility of this primarily in Chap. 6.

For game developers and the workplaces in which they make their games, it is both easiest and most important to start with Scope 2 emissions, those emissions generated through direct consumption of electrical power and for which *no one else* is responsible mitigating. In January of 2020, I released a survey to try and find answers to a series of questions about energy use and climate attitudes in game developers. Asking approximately 45 questions, the survey requested that developers share whatever data they had or could access about the energy consumption of their particular workplace, followed with some questions about their and their perception of their colleagues' concern about climate change. The most important question was how much electrical energy and gas (scope 2 and scope 1 emissions respectively) a given game studios uses. I asked developers: "Approximately how much electricity did your workplace use or pay for in its most recent billing period?" I received 30 responses, half from Australia. The survey also reached a wide variety of respondents from other counties around the world: France, the United States, Canada, Denmark, Germany, and Colombia all appeared in the survey. Respondents were given the option to remain anonymous or be identified, as well as the option of whether to identify their workplace or not. The survey adopted categories for size taken from the annual GDC State of the Industry survey, (GDC, 2020) one of the most comprehensive surveys of the games industry worldwide, with options for the number of employees at a respondent's workplace divided into the following ranges: 1, 2–5, 6–10, 11–20, 21–50, 51–100, 101–250, 251–500, 501+. Respondents worked at a range of different sizes of studio, from solo developers working out of their own home or in a shared co-working space, all the way up to studios with 500+ employees. Much like the spread of responses to the State of the Industry survey, the distribution was clustered at both ends of the range, with many solo developers and teams of less than 10 people, and a handful of large studios with multiple hundreds of developers.

In addition to asking about electricity and gas usage, the survey also asked whether respondents thought the data was being recorded more for financial or environmental reasons, the overwhelming response from survey respondents being that it was for financial reasons—only 10% (3 respondents) indicated it was collected more for environmental reasons. The survey also asked respondents to share what, if any, policies their workplace had in place to reduce its carbon emissions, including other green-friendly initiatives outside the sphere of direct energy consumption, which proved to be a surprising source of a number of actions already being taken.

Respondents were asked to rate their level of concern about the threat of climate change on a 1–5 scale from "not at all concerned" to "very concerned" as well as offer an assessment of their colleague's level of concern. Almost uniformly, respondents reported themselves at the high-end of the scale for concern about the threat presented by climate change, and higher than their perceptions of their colleagues' level of concern. This suggests respondents were perhaps individuals typically more concerned than the average game developer, reflecting the type of person more likely to self-select into completing the survey.

The overall picture painted by the admittedly small initial results, however, is striking, showing an industry that remains (with a few notable exceptions) largely unaware or uncertain about the extent of its own carbon footprint. Seventy percent of respondents did not think their workplace collected data about its energy use, or were unsure about whether such data was collected, and only seven out of thirty responses managed to provide concrete numbers for energy use. Developers were given the option of sharing their energy use data in whatever form available to them, with some sharing the dollar value cost of their electricity bill and some sharing the kWh measured by their power company and displayed on their bill. Wherever possible I have included best estimations for cost (in AUD) and kWh values. Cost is both useful to have as it offers an insight into local power prices, which may impact the feasibility of switching to renewables or offsetting Scope 2 emissions. Wherever only a dollar value was shared, I have attempted a best-guess at the quantity (in kWh) of power consumed using what available data I could find about energy prices for that particular location. From this I have attempted to arrive at a rough estimate of the carbon intensity associated with each energy consumption figure using a 'GHG equivalent' emissions figure calculated from the kWh of power generated/consumed multiplied by the 'emissions factor' (that is, the emissions intensity of 1 kWh of power generation), based on what emission factor data I could find for the given country or region the studio is located in.

The majority of this data was obtained from smaller studios, with disappointingly no data directly from companies in the 500+ category. It is telling that the two largest studios with responses to the survey, both of which were able to provide quite detailed data, are led by individuals keenly aware of the need to take climate action, as we shall see. The survey may have had more success in collecting energy use or emissions data if it were able to reach the upper-echelons of decision makers at game studios or even approached them directly. Often it was these company leaders who were able to provide the most detailed information and data on their company's energy use.

## SOLO DEVELOPERS

Beginning at the lower end of the scale, three solo developers offered a range of data points about their energy use. The lowest figure given in the survey was "$50" for a particular billing period (almost certainly one month) which, assuming a 28c kWh rate for their location in the Australian state of Victoria would mean somewhere in the vicinity of 178kwh worth of power used over the period (Sustainability Vic, 2020). This is roughly the same as the average Australian one-to-two person household power bill, reflecting the scale of operation of the respondent who is primarily a contract-based game audio worker.

Other responses from individual developers were typically higher: Brisbane based developer Marben offered a figure of '$140 (AUD) or 500 kWh for one month', and Seattle based Sketch House Games gave the figure of 768 kWh per month. According to the US Bureau of Labour Statistics, the average price of power in the Seattle area in Dec 2019 was just 10c/kwh, slightly lower than the US average, so we can estimate the overall cost to be somewhere around $77 USD (excluding supply charges) or approximately $115 AUD. Given that many solo developers work out of a home office, and the timing of the survey gathering responses during respectively, the hottest part of the year for Australian respondents, and the coldest for those in the Northern Hemisphere, these figures are probably at the upper end of seasonal trends, with heating/cooling potentially a large contributing factor for each. Still, we get a sense from this data that, perhaps predictably, game development done out of home offices seems to result in slightly higher power bills and more energy use than a similar household, a fairly expected finding and which may have some relevance for future studies that aim to account for the greater amount of work-from home now occurring post-pandemic. This also represents a challenge to Scope 2 accounting emissions as these carbon costs will likely now be attributed to the individual household rather than the studio according to existing GHG protocols.

## SMALL TEAMS

Fewer responses were gathered from small teams, but these responses still offered some insight. One respondent in a small team worked out of a co-working space, and so did not have access to their own energy usage data. This raises a similar challenge for accounting as the work-from-home developer, both of whom will struggle to disaggregate work usage from private power consumption. A game made from a team working in a co-working space therefore may have limited access to metrics about their energy use and emissions, even if technically the co-working space has responsibility for these. Comparisons to similar sized teams in this study may help provide an approximation, with the proviso that once again identifying the boundaries of Scope 2 emissions here can be challenging.

More useful are the figures provided by developers working out of small offices. Secret Lab, a team of 2–5 developers in Hobart, in Australia's southernmost state of Tasmania, reported a figure of '436 kWh over 3 months' which is a surprisingly small amount, only 145 kWh a month, quite a lot less than the highest of our solo-developers reported above especially considering it covers more than one employee. While it might not be the lowest overall power usage to appear in this survey, it might still be the cheapest thanks to Tasmania's abundance of hydroelectricity (only about 15c per kWh according to some source). This represents a cost of a little over $20 a month (excluding supply charges). Being a small team, in a regional capital city of Australia, they note that entire company's sources of energy consumption are 'lighting and computing equipment.'

They do raise, however, an important factor determining their overall carbon footprint, stating that 'being in Tasmania, we're required to fly quite a bit – either to the mainland for Australian events, or overseas for international events.' The issue of business flights is one that came up a number of times, raised by several respondents. Travel emissions are typically accounted for as a Scope 3 emission—as they are another business generating them. Nevertheless, they are an important emission to offset, and conventionally considered part of a business' responsibility due to their outsize impact on overall footprints. We will return to the question of flights and their impact on the carbon footprints towards the end of this chapter. In terms of the quantity of Scope 2 emissions, the Hobart based team and their figure of 145 kWh in a month is still a decrease from most of the figures provided by the solo developers whose energy usage (despite being only one person in each case) can be attributed to their work energy consumption and its aggregation with domestic power usage. For the game development workplace that only has lights and IT equipment, most likely no one is having hot showers at the office, necessitating heating of copious amounts of hot water, for instance.

Another respondent from a small studio of 6–10 people working in Western Australia, provided a figure for their power costs totaling $297 for one bill. The typical billing period in Perth, where this studio is located, bills around a 60-day period. The network operator, Synergy, has a rather high supply charge of about $1 per day, and charges a fairly pricey 28c kWh (the state of Western Australia is not connected to the rest of the Australian energy system due to its distance from the other states). This means that about $60 of this bill is supply charge, the remaining $237 representing a figure of approximately 846 kWh of power usage, or 423 kWh a month, again over the peak summer periods of Dec 2019 to Jan 2020.

For both locations—Hobart and Perth—we can take the Scope 2 energy consumption figures and multiply them by the respective Greenhouse Gas emission factors for the region. The Australian government provides emissions factors for consumption of electricity purchased from the grid in all states (Dept of the Environment and Energy, 2019) as do certain electricity retailers. For Tasmania the emissions factor provided is 0.15, and for the part of Western Australia centered on Perth the factor is 0.70. That gives us the following emissions figures for a single month of small team development in both Hobart and Perth:

Hobart—145 kWh × 0.15 = 21.75 kg of $CO^2$ equivalent emissions.
Perth—423 kWh × 0.70 = 296 kg of $CO^2$ equivalent emissions.

The absence of domestic energy consumption from the figures of small teams working in dedicated office spaces helps illuminate the challenge involved in accurate accounting for solo developers and work-from-home employees. The huge explosion in both of solo and small teams all across the world can be seen in the GDC State of the Industry survey, (GDC, 2020) and for these the disaggregation of work and home consumption may prove difficult. We also should not rest on the fact that small teams are using small amounts of power. As shown by the contrast between the emissions factor difference between Tasmania and Western Australia, small amounts of power in high emissions intensity regions add up to substantial emissions in places with high emission factors for their power generation. That work from home employees are unlikely to be able to disaggregate their work power usage from their private consumption only further underscores the argument I make in the rest of this book about the urgency of grid-scale decarbonization. One conclusion from this is that perhaps game developers and the industry at large might need to contemplate action beyond its immediate sphere of influence. Games developers who wish to start thinking ecologically, make ecological games, and act on the recognition that their industry is not separate from the climate crisis—may need to act outside their immediate sphere of influence. In this way the games industry itself is given political charge by the issue of its emissions, its work organization, and may need to begin to expand the scope of its actions.

## LARGE STUDIOS

Two final respondents to the survey provide perhaps the most illustrative data yet around the scale of energy consumed by game development teams, despite the fact that even the biggest of these is as much as two orders of magnitude smaller than parts of the Triple-A industry. The first is Wargaming.net's Sydney office, which employs approximately 100 people, occupying a whole floor of a large commercial office space in the city's CBD. Chief Technology Officer Simon Hayes provided figures personally in the survey, as well as offering some background and further details in a follow-up interview in which we discussed the studio's carbon footprint and energy use, and some intimidating figures. The second was Nicholas Walker of Space Ape games, a London based mobile game developer, employing around 250 people, who provided kWh data both in the survey and a number of reports the studio has published to its website (Space Ape Games, 2020). These reports demonstrate a proactive approach to measuring energy intensity and carbon offsetting activities, a definitive climate leader in the industry. Both Wargaming.net's Sydney office and Space Ape Games studio in London provided information on further emissions mitigating strategies beyond their energy consumption as well. Together with the data provided in the survey responses, they make it possible to paint a clearer picture of the scale and intensity of energy use in modern large-scale game development as well as what can already be done about it right now.

Wargaming.net's Sydney office reported a power usage figure of 400,000 kWh of electricity per annum, or about 33,000 kWh a month. If Wargaming.net are paying a state average rate of 27c kWh that amounts to about $9000 monthly electricity bill. In an encouraging development, Hayes reported that the 400,000 kWh figure is also 100% carbon offset, and I asked him how they achieved that, what was involved, and whether it was a difficult change to make. Hayes confessed to being a lifelong "greenie", who has solar panels on his home, and has been privately purchasing renewable energy for close to a decade, highly conscious of environmental issues personally. However he also explained that the decision to purchase renewable or 100% carbon offset energy was driven by the team itself. Hayes explained that the view of the management team was such that being a company that is a 'force for good' is a successful corporate talent strategy, helping them to retain their best staff. Hayes flagged other policies in a similar vein, like supporting diversity initiatives in the workplace, mentioning staff are given volunteer days to use in the community, and that as a company they decided to use one of these days to attend the September 2019 climate rallies. Hayes mentioned that they made T-shirts for staff with the Wargaming.net logo, with a message on the back: "Save the planet! That's where I keep my videogames!" A lot of the social and environmental initiatives are also staff driven, with small teams given decision making power to choose and direct strategies to improve the company's environmental footprint. These range from relatively minor changes, like choosing which type of paper to use in printers and photocopiers in the office, to switching to buying bulk 1 liter containers of yoghurt rather than smaller individual pots to cut down on packaging waste, as well as outfitting the kitchen with reusable plates and cutlery. The team also made the choice to switch over to 100% renewable energy, and rather than being difficult to implement, Hayes reported that the switch was essentially the same as changing domestic power providers. All it involved was finding a new company and signing up—no power purchase agreements needed to be arranged, and there were not even any additional costs. An unexpected upside to changing power providers was the discovery that they had been paying rates that were higher the necessary. Hayes explained that the studio's electricity bill 'actually went down' as a result. Choosing the power company Meridian Energy, the largest renewable generator in Australia, meant that the power used by Wargaming. net Sydney office was now 100% generated by renewables, and it cost them nothing extra.

The other great success story is from Space Ape Games, the London based developer of popular mobile games like *Transformers: Earth Wars* and *Rival Kingdoms* (*Space Ape Games* 2015, 2016). Space Ape reported that in 2018 they consumed around 150,000 kWh of electrical energy and 95,000 kWh (equivalent) of gas—for a total of only around 20,500 kWh a month, despite being a larger studio (up to 250 people). Like Wargaming. net's Sydney office, they also set a goal of being carbon neutral, and according to their green reporting site, has achieved it as of September 2019. This online report, a fantastic resource for the games industry, outlines the steps they took and provides a four step process for other developers to follow to achieve similar carbon neutrality.

The report focuses on two areas of emissions that would be considered Scope 2: office energy use, and transport including flights and hotels. It also captures many emissions considered Scope 3—i.e. emissions made by others but generated on behalf of or through the use of Space Ape Games products and services. This includes cloud computing (Scope 3 if provisioned by external companies/data centers, Scope 2 if on premises), deliveries and purchases, and "mobile device usage by our players"—this last being perhaps the most surprising and ambitious part of this project which, as far as I am aware is the first attempt at actively accounting for and offsetting the emissions generated by players themselves. Accounting for and offsetting the emissions of this many players is a substantial task, one we return to in Chap. 6 when we discuss the emissions of playing games. For now we will stick to Scope 1 and 2 emissions, those attributable solely to the developers themselves. Curiously, Space Ape was also one of the only companies to report Scope 1 emissions through an amount of gas usage (most likely for heating), reporting the 95,000 kWh (equivalent) figure for 2018 mentioned earlier. Head of Technical Operations Nicholas Walker also noted in his response to the survey that 'we offset 2x our annual carbon footprint, [and] have a commitment to reduce studio footprint by 10% for 2020.'

Space Ape are clearly achieving much already, accounting and offsetting emissions not just generated by themselves (scope 1 and 2 emissions) but also for the emissions embodied in new hardware acquired by the company, noting in their online report: 'We decided to focus on our tech purchases for the year... We found that Apple publishes a lot of very helpful info about their products to assist in carbon footprint calculations. The same can't be said of Dell, HP, Samsung and others... Nevertheless, there was enough data out there to let us make the necessary footprint estimates' (Space Ape Games, 2020). According to the GHG protocol, embodied emissions from manufacturing are 'owned' by Apple, Dell, HP, etc. however it is not yet commonplace to find manufacturers taking responsibility for offsetting these emissions themselves. Perhaps this may change with regulation, consumer expectations, or even other corporate actors looking to take action into their own hands and demand better disclosures, placing a premium on carbon neutral procurements. Walker added the following comments about the corporate goals of Space Ape in the survey:

> We are also evangelising sustainability within the industry through public speaking and through work with the UN Environment group, Playing for the Planet. On the smaller end of things we try to encourage sustainable behaviours ...We have a stash of reusable coffee cups for visits out to grab coffee, we have well labelled and numerous recycling bins, we source food for the office [sustainably] with reference to plastic containers, organic certifications, etc.

The scale of action currently being taken by industry leaders in games and sustainability, then, ranges from the macro, with emissions accounting and offsetting, to micro-decisions about green or sustainability signaling in workplaces and procurements. The final part of the games sector that has yet to be addressed, however, and entirely absent from the survey and these figures however, is the core of super-large Triple-A development studios. The elephant in the room in terms of sheer size, the mega studios of Triple-A game development is the gravitational center of much of game development today, and aside from the two studios above, did not meaningfully appear in the survey data. Even though a number of employees from these mega-studios completed the survey, none had access to the data I was seeking. In the next section I look to the reporting tools of these corporate mega-studios, in the form of corporate social responsibility (CSR) reports, which are becoming more common avenues of disclosure as calls for climate action increasingly focus on large institutional actors. These CSR reports regularly contain details about energy and emissions as well as other socially responsible activities, responding to public demands for corporations to act—and be seen to be acting—sustainability and in a socially responsible manner.

## TRIPLE A MEGA-STUDIOS

Many of the larger publicly owned game companies now publish annual CSR reports, from companies like Nintendo, Microsoft, SONY, Ubisoft, EA and Rovio—though each varies greatly in the level of detail provided. These reports follow similar corporate trends toward disclosure of energy use and carbon emissions, providing a wealth of heretofore untapped information about the emissions intensity of game development for what is certainly the largest part of the games industry –in terms of employees, revenue, player numbers, and visibility. A brief summary of the relevant data from each of these companies as well as the three main console platform holders follows, with a table at the end collating all the available information on emissions gleaned from both the survey and the CSR reports.

## Rovio

The Finish based developer of the *Angry Birds* games, Rovio, provides an annual CSR report that offers both future emissions targets and concrete data about current emissions and offsetting activities:

> The key performance indicators for environmental responsibility are the proportion of Rovio's cloud service providers using renewable energy in their electricity consumption (target: 100%), as well as committing to offset the $CO^2$ emissions arising from company air travel and the use of office premises. (Rovio, 2020: 11)

The report helpfully provides figures for office energy use of 0.53 tonnes of $CO^2$ emissions per person, and carbon emissions from business associated air travel (1.64 tonnes per person). The report notes that this represents total emissions of 1013 tons of $CO^2$ equivalent which was 'fully compensated through the United Nations climate carbon offset platform' (Rovio, 2020: 11). As we will see, this per-employee emissions figure is comparatively low, and an encouraging figure for Scope 1 and 2 emissions. That flight emissions amount to triple office energy emissions is also telling, though difficult to extrapolate from. The report further claims to have gathered data from cloud service providers which suggest 'renewable energy accounted for approximately 63% of the electricity used for Rovio's cloud services in 2018' though makes no mention is made of how this figure is trending. The overall picture here however is quite encouraging. While there is no date attached to the target for 100 renewable cloud provider services, and there are still good reasons to prefer emissions reductions over offsets after-the-fact, to some degree offsetting is going to be a necessary part of the carbon neutral future for the games industry and is better than nothing. Even this much progress on achieving a carbon neutral business is still a significant achievement. The relatively low emissions per employee figure (barring flight emissions) may reflect the advantage that mobile game developers have, in that the general computational intensity of mobile game development is lower overall. We certainly see this in the emissions offsetting of player emissions (covered in Chap. 6) compared to console and PC gaming. To the best of my knowledge only Space Ape Games and Rovio have yet been able to account and offset player emissions. Mobile game developers may well face an easier task in achieving Scope 3 carbon neutrality than developers of games for other platforms.

### *Ubisoft*

One of the largest game developers and publishers in the world, Ubisoft on its own represents a substantial slice of global game development, with major centers of development in North America, Europe and Asia. As a publicly listed company on the French 'Euronext' exchange with subsidiaries around the world, Ubisoft publishes detailed annual reports with a fairly comprehensive CSR component. Ubisoft was one of the first major signatories of the UN Playing for the Planet initiative in late 2019, and provides detailed data for emissions and energy use in its 2020 report, with figures that dwarf any discussed thus far. In 2019, the company reported energy use of 51,287,000 kWh of electricity—equivalent to only 9510 tons of $CO^2$ emissions, and a relatively modest annual per employee figure of 2776 kWh (Ubisoft, 2020: 170). The combination of these figures suggests that Ubisoft had around 18,400 employees in 2019, matching other estimates. Knowing these numbers allows us to arrive at a $CO^2$ emissions figure of 0.51 tonnes per employee, a figure extraordinarily close to that provided by Rovio's office emissions. It seems to be the case that Ubisoft achieved this by purchasing '66.7% of the electricity consumed in 2019 across the Group's sites...from renewable sources' (Ubisoft, 2020: 170). They provide additional break down of this mix, noting that its studios based in Montreal and the rest of Quebec, 'are powered by the energy supplier Hydro-Québec, 99% of whose electricity is generated using hydro-electric dams' (Ubisoft, 2020: 170). Clearly, where a studio gets its power from has a large impact on its carbon footprint, with proximity to plentiful hydro-electric power sources producing a much lower per-employee figure than otherwise.

## *Electronic Arts*

Initially, very little data was contained in Electronic Arts' (2019) CSR report, with the entire document covering barely two pages in the larger annual report and including no hard figures or evidence to support broad claims of environmental responsibility (EA, 2019: 5–6). In 2020, however, the company released a dedicated impact report that greatly increased the amount of information disclosed, including figures for companywide energy use and a number of actions being undertaken to reduce EA's environmental impact. These include 'reducing our carbon footprint in the delivery of games and services,' increasing power and water use efficiency in data centers, and making 'environmentally-conscious choices in our offices worldwide' (EA, 2020: 29). In an infographic, the report claimed that in 2020 EA used 71.49 MWh (or 71,490 kWh) of energy in North America, 18.26 MWh (or 18,260 kWh) in Europe, and 3.56 MWh (or 3560 kWh) in the Asia-Pacific region—for a total of 93.31 MWh (or 93,310 kWh) of energy as part of its global operations. This is an *extremely* low figure given the stated count of 9800 employees and over 3000 square feet worth of facilities (EA, 2020: 32)—what I strongly suspect is that this is a typographic error. This impression is reinforced by the fact that the very next page of the report states the company has made *reductions* on the order of 829,000 kWh per annum (or 0.829 Gwh) 'from LED light retrofits, voltage harmonizer, and automated light sweeps' (EA, 2020: 33) which would mean they reduced their overall power by eight times their total consumption, which seems improbable at best. Assuming the earlier totals should really be listed as gigawatt hours not megawatt hours (in other words, that they are 1000 times larger) makes everything in the report make a lot more sense, and puts their per employee per annum figure at least in the same order of magnitude as other companies.

Disappointingly, despite these figures come with no associated carbon emissions attached to this energy use, leaving us to estimate Scope 2 emissions based on broad regional emissions factors. Applying country-level emissions factors (and assuming that the above MWh figures should be the larger GWh) we get approximately 37,000 tonnes of $CO^2$ equivalent emissions for 2020. This works out to be around 3.8 tonnes of $CO^2$ per employee per annum, which is in line with much of the rest of the Triple A industry.

The report notes that 'two of our key cloud providers power their data centers with over 50 percent renewable energy' (EA, 2020: 31) though it's not clear if EA considers this within Scope 2 or 3 emissions—onsite data centers being Scope 2, offsite third party data centers being Scope 3. The report also notes however that 'many of our EA-Managed data centers are on track to be powered by 100 percent renewable energy by 2026' (EA, 2020: 31). The complexity in measuring emissions intensity across the whole energy-mix for a global company is certainly not without its challenges, however greater accuracy is more easily achieved if it is done by those inside the company with directly access to the data. It is greatly desirable to have companies take on this work themselves, as it will always be less reliable when it has to be established by outside estimates.

It is important to recognize just how much of an improvement the 2020 'impact report' is over previous offerings, with the 2019 annual report only including a boilerplate paragraph on the importance of sustainability to the company:

> We aim to integrate environmental responsibility and sustainability into our operational and product strategies. We reduce our carbon footprint by the manner through which we bring our games and services to players and by making environmentally-conscious choices in our offices worldwide. (EA, 2019: 6)

The change in willingness to disclose energy used documented in the 2020 report and the seriousness of some of the commitments EA has made is notable, seeming to reflect a shift in wider corporate attitudes towards environmental issues which are increasingly seen as affecting companies social license to operate (Gunningham et al., 2004; Provasnek et al., 2017). EA still appears to be primarily relying on their carbon footprint reducing due to the shift from physical media to digital games distribution. We will see later in the next chapter, however, that digitization alone, while a necessary step, is not a sufficient response, and in many instances results in a shift in responsibility for emissions either onto third parties and consumers. For this reason it is encouraging to see data center energy use considered in the report, and the goal of 100% renewable powered data centers, however there is no mention yet of a target date for complete decarbonization of the business. In future reports the disclosure of GHG emissions as well as energy consumption would be a substantial improvement as it remains essential to measure and drive change towards a carbon natural games industry.

The comparison between EA and Ubisoft—two Triple-A developers with quite different per-employee emissions figures—shows that there is a wide range of intensities across game development, even in the Triple-A development space. Despite having almost double the employees of EA, because of its geographical access to renewable hydro-electric power, Ubisoft has about 1/7th of the per-head emissions intensity. Game developers wishing to rapidly curtail their emissions will clearly be well served by focusing on where they purchase power from, and investigating power purchase agreements that might help to drive further investment in renewables.

### EA Dice

EA Dice, makers of the *Battlefield* series of games, based in Sweden, provides a similar CSR report that has some details about its sustainability practices. While it begins with and repeats the same boilerplate paragraph from the general EA 2019 reporting document, it also adds more details about local operations and activities to increase sustainability and lower emissions. These include installing LED lightbulbs, using green cleaning products, compostable food and beverage containers, recycling programs, and, perhaps most importantly, a renewable power purchase agreement:

> At DICE, we also have a green electricity agreement with our current provider, meaning that all electricity is generated through renewable energy sources; wind and/or hydroelectric power. (EA DICE, n.d.: 5)

No figures are provided for how much electricity was consumed by EA Dice, nevertheless it is reassuring that the energy used is carbon neutral, an encouraging development that ought to be celebrated. The use of fully renewable energy, once again, underscores the importance of grid-scale transformation and the imperfect but essential aim of sending price signals to energy markets that emissions intensive power generation will no longer be tolerated, with large corporations having a disproportionate influence in this area. As we saw in the case of Wargaming's Sydney office earlier, sending this signal seems to no longer involve the financial penalties that it once did. Renewables are increasingly the cheapest option for new generation, and in an increasing number of countries are beginning to displace older, more emissions intense forms of generation (IEA, 2020).

## PUBLISHERS AND PLATFORM HOLDERS

### Nintendo

After receiving some high-profile negative attention in past years, scoring dead last in an assessment of the green credentials of the digital games industry by Greenpeace in both 2010 and 2015, Nintendo appears to have cleaned up its act somewhat. In its 2019 CSR report, Nintendo (2020) disclosed the amount of energy used across the various regions it operates, with a global figure of 895,731.2 kWh of power used from entirely renewable source and 23,215.7 kWh from non-renewables, a split of about 98% renewable energy to not. The report also provides a snapshot of the size of the company, with a figure of global permanent employees at 4548 (plus 173 temp workers) and providing the low energy use figure of 202.1 kWh per annum, per employee. According to Nintendo's calculations, this amounts to 1.6 tonnes of $CO^2$ emissions per employee—an annual emissions intensity for the company of about 7554 tonnes of $CO^2$ equivalent. It is important to note, however, that this figure almost certainly excludes the manufacturing of Nintendo consoles and hardware, as these are typically subcontracted out to manufacturers, placing the emissions for the assembly of devices outside the GHG protocol's Scope 1 and 2 emissions. For the Nintendo Switch, these emissions are probably attributable to Foxconn in Taiwan and Hosiden in Japan (Lee & Tsai, 2017). Some may find this arms-length arrangement unsatisfactory, a product more of the structure of accounting rules than actual responsible attribution. Similarly, data center power usage and emissions intensity involved in running the Nintendo digital store, multiplayer networks and other online services may well be outside the boundaries of Nintendo's reportable emissions. If not, and regardless of whether or not we agree with these various boundaries, as a result it is quite likely that these figures above largely represent game development (and game publishing) activity for Nintendo, allowing us to use them to contribute to the emerging picture of the intensity of game development, if somewhat leaving the picture of hardware manufacturing rather obscured. The other platform holders, Sony and Microsoft, disclose more information about hardware, but in general this aspect is beyond the scope of this chapter. I return to this issue in more detail in Chap. 7, where the 'periodic table of torture' shows just how varied and complicated the environmental harms associated with gaming hardware are, encompassing far more than just carbon emissions.

### *Sony*

In contrast to the picture of Nintendo's corporate footprint, Sony's status as a multinational consumer electronics manufacturer with trillions of dollars of revenue and large factories of its own, makes it somewhat less useful for determining game development intensity despite a very thorough CSR. The method that Sony uses to publish its data in the report also does not appear to provide hard numbers, only breakdowns for the different arms of the business and end consumer emissions. Particularly unhelpful for our purposes is the fact that it breaks down its emissions intensity reporting according to geographic region rather than business category, so we only know that Japan/East Asia accounts for a majority (approximately 1.1 m) of the company's 1.38 m tonnes of Scope 1 and 2 emissions in 2019. Much of this energy is from factories that produce consumer electronics, but without more detail about each business arm we are left without any way of estimating the intensity of the game portion of Sony's business. We do know, however, that there is a similar arrangement to Nintendo, with Foxconn manufacturing PlayStation 4's for Sony (Mishkin, 2013). The report notes, however, a *very* modest aim to 'request manufacturing outsourcing contractors with large business transactions to monitor GHG emissions and reduce GHG intensity by 1% per year.' Needless to say, 1% per year is almost certainly inadequate, and the framing of it as a 'request' leaves it unclear how enforceable this might be, how, or even if it will be measured. The one upside though, given Sony's impressive size, is that any initiatives could have substantial impact if taken seriously and followed through on. It is still a very far cry from our goal of carbon neutrality however.

What we find in Sony's CSR report instead, however, is an overall energy consumption figure in terajoules (24,400 TJ) (Sony CSR, 2020: 109) which works out to 6,777,777,777 kWh (or 6.77 TWh). We are unable to reckon, however, how much of this is related to game development or associated with game publishing, the PlayStation network, or anything else really, being bundled together with the rest of Sony's corporate energy use. It does note, however, that 500 TJ (0.1389 TWh) of renewable energy was used, and that 'of the electricity used at Sony worldwide, electricity generated by renewable energy accounted for approximately 5%' (Sony CSR, 2020: 127)—a rather disappointingly low proportion. The rest of its energy usage reporting takes the form of case studies and anecdotes of small movements towards carbon neutrality that it is making, but it does at least have an overarching goal of total carbon neutral operations by 2040 (SONY CSR, 2020: 126). The report also notes the increasing utilization of renewable energy in Japan and discusses green power certificates and the search for direct power purchase agreements (SONY CSR, 2020: 127). The report also notes that in Europe Sony has two sites powered by rooftop solar installations and purchased renewable certificates to achieve '100% renewable energy usage' (SONY CSR, 2020: 128). One of these is the Thalgau plant operated by subsidiary Sony DADC Europe, which manufactures all PS4 and presumably now PS5 Blu-rays (Aslan, 2020: 160). Sadly, this is the extent of the information provided, as it does not mention any Sony first-party game studios like Guerrilla Games or Media Molecule in Europe, or North American studios like Santa Monica Studio, Insomniac Games, or Naughty Dog. I was unable to find any mention in the Sony global CSR report of any specific game studios. Though they may be owned by Sony, it's possible that they remain separate corporate entities not counted within Sony's corporate emissions boundary, or are simply aggregated within the larger figures. More work investigation is need to establish whether these studios are, or plan to be, bound to Sony's overall 2040 carbon neutral target.

Data centers owned by other companies around the world are, again, outside the scope of Sony's emissions reporting boundaries, with the report only noting that 'Sony's environmental mid-term targets include the target of prioritizing the use of energy-efficient data centers' (SONY CSR, 2020: 124). As a target, however, this leaves a bit to be desired, with no numbers or dates for this to be achieved, and currently insufficient detail at present to estimate data center energy usage or emissions.

The Sony CSR report's focus is more on the emissions for the entire global company, however it does include estimates of downstream consumer (or Scope 3) emissions, noting that 'the amount of greenhouse gas emissions from Sony's overall value chain in fiscal 2019 is estimated to be approximately 16.24 million tons' (Sony, 2020: 108). The largest volume of these emissions, approximately 9.57 million tons, was from 'energy consumed during product use.' This is the Scope 3 emissions of end-users switching on Sony's TVs, PlayStations and other electronic devices. Even excluding these downstream emissions, as mentioned earlier, Sony's overall Scope 1 and 2 emissions totals to around 1.3 million tonnes—*several* orders of magnitude larger than even the biggest game development studio included in this chapter, and the largest per-employee annual emissions intensity figure at about 12.35 tonnes, reflecting the quite different nature of the Sony business.

## *Microsoft*

Much like Sony and Nintendo, Microsoft is a multinational company with corporate interests in both the software and hardware aspects of game development. Microsoft owns the Xbox family of consoles and the Windows operating system, still the most popular PC gaming platform, particularly for high-performance gaming, streaming, and the platform on which most game development is likely taking place. Microsoft has also become a global leader in sustainability initiatives, publishing detailed emissions breakdowns for the entire global corporation for several years now, and has a number of top-level positive achievements and some pf the most ambitious climate goals. It purchases 100% renewable electricity, quite credibly claiming to be one of the largest purchasers of it in the United States, (Microsoft, 2020a: 33) its data centers and entire global operations are also made carbon neutral through offsetting mechanisms, (Microsoft, 2017: 2; c) it has a target of being carbon negative by 2030, and an ambitious target of removing all of the emissions generated by the company by 2050. As part of this detailed corporate reporting, Microsoft regularly discloses an impressive array of facts and figures about its environmental impact and progress towards targets. The 2020 Environmental Sustainability Report includes an appendix with Scope 1, 2 and 3 emissions for the company, broken down into emissions sources like Scope 2 location-based emissions (4.1 m tonnes of $CO^2$ equivalent emissions) and even a host of Scope 3 emissions that includes employee commuting (317,000 tonnes), 'use of sold products' (3.025 m tonnes) and others. The total carbon footprint of the global corporation is tallied up to be around 11.164 m tonnes of $CO^2$ equivalent emissions (Microsoft, 2020d). Summarizing the company's approach to carbon offsetting and neutrality, the report notes the following approaches, which are useful guidelines for anyone looking to do similarly:

> Our approach to renewable energy has two core tenets: **regional impact** and **additionality**. We have focused on regional matching to operations, because where and how you buy matters – the closer the new wind or solar farm is to your datacentre, the more likely it is those zero carbon electrons are powering it. Microsoft, as a result, is sometimes a market driver, striking the first or the largest corporate [power purchase agreements] in a state or region that was not previously viewed as a good market for renewables. We also focus on additionality, using our capital to fund new projects that may not succeed without our investment. (Microsoft, 2020d: 18)

The importance of additionality in carbon offsetting initiatives is reflected in other reports on the weaknesses of many carbon offsetting schemes, and represents a welcome focus (Compensate, 2021). One of the ways that Microsoft has made such progress in this area has been through the introduction of an internal carbon price:

> We established our internal carbon fee in 2012 to fund our carbon neutrality commitment. In 2019, we raised the fee to $15 per ton, which we charged to each business group across Microsoft based on their Scope 1 and 2 carbon emissions and business air travel. (Microsoft, 2020d: 19)

The data provided in two main documents– the devices sustainability report and environmental data fact sheet—does not provide sufficient data on the energy used by specific business groups, making it difficult to gather much in the way of comparable data on the energy use and the emissions intensity of the gaming divisions of the company. Manufacturing agreements similar to both Nintendo and Sony's appear to have existed for earlier Microsoft consoles—with Flextronics a Singaporean and American manufacturer delivering reportedly 90% of the Xbox One with Foxconn manufacturing the remainder (Lee et al., 2013). Even here, Microsoft seems to acknowledge the need for progress, with the report noting that 'in 2020, we updated our Supplier Code of Conduct to now require a greenhouse gas emission disclosure' (Microsoft, 2020d: 19) which could cover companies manufacturing and supplying the devices on behalf of Microsoft.

I have not been able to find confirmation that the Xbox Live online service is hosted on Microsoft's own data center network, however it would stand to reason that it would, and if confirmed represent to my knowledge the only gaming network that is fully carbon offset. Microsoft's Xbox Game Studios subsidiary, which owns studios such as 343 Industries (developing the *Halo* series of games), Doublefine, and even Mojang (the developers of *Minecraft* which Microsoft bought in 2014), may also be within the parent company's CSR emissions reporting boundary and thus be covered by its carbon neutrality and offsetting initiatives. Verification of this would be greatly desirable, however it is noteworthy that the gaming part of the company does not frequently appear in Microsoft's sustainability reporting. One of the few mentions that do appear is the following note about a low power mode added to Xbox consoles:

> Xbox recently added Regulatory Standby Plus (RS+) as a new power mode. At a high level, RS+ will provide the power savings and environmental benefits of RS with the benefits of keeping the user's OS and content up to date. This can reduce power from 15W to less than 2W during standby mode. (Microsoft, 2020d: 19)

The power consumed by gaming devices is an important part of the industry's footprint (discussed in Chap. 6) however it's not specified in the report which Xbox device or devices these new modes apply to, and similar confusion has arisen before. Overall, the impression that one gets is that the games group of the company might not have quite the same dedication to sustainability as the rest of the business. In August of 2020 a news story broke, first picked up via a forum thread discussing a then-new report into the production of a series of 'carbon neutral' Xbox consoles. For whatever reason, the Microsoft authored report included a picture of the then still in-development Xbox Series X console. Because of this, new reports reasonably concluded that these brand new consoles were being made sustainably. Upon further investigation, however, it turned out that the carbon neutral certification achieved only referred to Xbox One X consoles, and as reported by Ryan (2020) the image of the next-gen console was eventually removed from the report. Despite this mix-up, it is still a welcome sign, despite perhaps the conclusion that the newest consoles are *not* being produced to this same carbon neutral standard. The Microsoft 2020 Devices Sustainability report provides the following explanation for what was involved in achieving this certification, not a simple process:

> To achieve CarbonNeutral® certification, we worked with an independent third-party to conduct a lifecycle assessment (LCA) of the greenhouse gas emissions of the Xbox console – from the raw materials, manufacture and distribution to the use and disposal of the console, its controllers and its packaging. It is estimated to save about 616,000 tons of $CO^2e$ a year. (Microsoft, 2020b: 51)

The report also describes it as a 'pilot' (Microsoft, 2020b: 51) but does not specify what (if any) parts of the initiative are to be taken up in the Xbox Series X console. Nevertheless, given that most hardware manufacturing of games consoles is done by third party manufacturers under contract, and thus not usually accounted within the boundaries specified by the GHG protocol (i.e. not attributable to either Sony, Nintendo or Microsoft) even this small bit of attention paid to the costs of manufacturing gaming devices, and the life-cycle emissions they produce is a welcome development.

## TRAVEL AND FLIGHTS

This chapter has so far only touched in passing on one of the most carbon intensive parts of all game development related business activity, which is the numerous flights involved in the increasingly international, collaborative and distributed nature of development. Not only are international flights between different studios and publishers now an expected part of game development, but so is the travel of members of the development team to different parts of the world for press events, conferences, and other launch activities. The Rovio CSR report mentioned earlier provides some regarding business flights and emissions intensity, with emissions from flights being *triple* the emissions of regular office-based energy use. This will be a crucial area for both future researchers and business emissions accounting to acknowledge and account for. The GHG Protocol suggests some basic arithmetic for accounting for Scope 3 business travel (including flights) based on number of employees in a travel group, multiplied by the distance travelled, multiplied by the emissions factor of the transport method (See Chap. 6 in WBCSD, WRI, 2004). These emissions factors are often available directly from the airline or travel company itself. Simon Hayes of Wargaming.net noted in our conversation that a flight emissions logging system was still in development for their organization, already on their radar as an important element to capture: 'we still need to address flights and things like that, and address transport that people use to get to work.' In the Space Ape games spreadsheet is logged around 500 flights and hotel stays, totaling around 177 tonnes of $CO^2$ equivalent emissions, from around 1.7 million kilometers worth of flights. Microsoft also logged 329,356 tonnes of $CO^2$ emissions from its business flights, though again what if any portion is related to game development is impossible to say. Together, these Scope 3 additions to the already quite substantial Scope 1 and 2 emissions adds up to quite an emissions intensive picture for just one industry. The last part of this chapter extrapolates from these figures to estimate the total emissions intensity of the modern games industry.

## Total Game Development Emissions in 2020

To arrive at a figure for the overall games industry and its emissions intensity for a given year, we can approach a reasonable estimate by combining the number of employees in the industry with what we estimate the emissions on a per-employee basis to be. (Fig 4.1) Unfortunately there is no single body that collects world-wide employment figures for the global games industry. Aphra Kerr (2017: 100) provides the next best thing, compiling a variety of national statistics and presenting them in a table in her book *Global Games*. Drawing on data mostly collected around the year 2015, Kerr (2017: 99) notes the difficulty that such an approach represents, with the following caveats applying.

> In some cases, the figures appear to understate the numbers employed as official statistical sources struggle to catch up with changing sectoral and occupational boundaries. It is likely these figures under-estimate the numbers of freelancers, part-time workers and others involved in the industry.

Similarly, and drawing on updated versions of quite a few of the same sources will face the same sort of challenges and risks similar under-estimation. Adding up Kerr's (2017) totals for the countries listed, and which by no means is the full extent of the global games industry, gives a figure of around 210,000—representing the employees of the United States, Canada, UK, Germany, France, Spain, Finland, Sweden, Poland, Netherlands, Ireland, Italy, Japan, China, South Korea and Australia for roughly around 2015. The years since Kerr's data was collated have been good to the games industry, seeing it growing at a substantial rate. The

| Studio | Location | Employees | Annual kWh (Scope 1&2) | Asociated CO2 equiv emissions p.a (tonnes) | OFFSET or RENEWABLE? | kWh per annum, per capita | CO2 per annum, per employee (tonnes) | Scope 3 Emissions (tCO2eq) if avialable |
|---|---|---|---|---|---|---|---|---|
| MarbenX | Brisbane, Aus | 1 | 6,000.00 | 4.80 | - | 6,000.00 | 4.80 | - |
| Maize Wallin Audio | Melbourne, Vic | 1 | 2,136.00 | 2.29 | - | 2,136.00 | 2.29 | - |
| Sketch House Games | Seattle, USA | 1 | 9,216.00 | 1.82 | - | 9,216.00 | 1.82 | - |
| Secret Lab | Hobart, Australia | 2-5 | 1,852.00 | 0.35 | 90% HYDRO, 10% WIND | 370.4 | 0.07 | - |
| Anonymous WA studio | Perth, Australia | 6-10 | 5,076.00 | 3.55 | - | 0.3552 | 0.04 | - |
| Wargaming.Net (Sydney) | Sydney, Australia | 51-100 | 400,000.00 | 328.00 | 100% RENEWABLE/OFFSET | 4,000.00 | 3.28 | - |
| Space Ape Games | London, UK | 101-250 | 245,000.00 | 44.70 | 100% OFFSET | 984 | 0.18 | 181.34 |
| Rovio | Espoo, Finland & Stocl | 466 | - | 248.00 | 100% OFFSET | - | 0.53 | 19,476 offset |
| EA DICE | Sweden | 640 | - | - | 100% RENEWABLE | - | - | - |
| Nintendo | Global (JPN, US, EU, / | 4,721 | 1,008,946.90 | 7,553.60 | 98% RENEWABLE | 202.1 | 1.60 | - |
| EA | Global | 9,800 | 93,310,000.00 | 37,280.03 | - | 7775833.333 | 3.80 | - |
| Ubisoft | Global (CAN, FR, ROM | 18,400 | 51,287,000.00 | 9,510.00 | 66% RENEWABLE | 2,776.00 | 0.52 | 29,080.00 |
| Sony (Global report, whc | Global | 111,700 | 6,666,670,000.00 | 1,380,000.00 | - | 59,683.71 | 12.35 | 9,570,000.00 |
| Microsoft (Global report, | Global | 163,000 | 10,244,377,000.00 | - | 0100% RENEWABLE + OFFSET | 62848.93865 | 0 | 6,715,261.00 |

**Fig. 4.1** Game developer emissions around the world. Full table online: https://bit.ly/3yGY7wT

latest data sees the US now employing 73,399 (Cook, 2021) up from 42,000, the UK employing 16,140 (up from 10,869), (UKIE, 2020) Spain increasing from 3380 to 7320 as well as '2,500 self-employed and 5,900 indirect jobs' (DEV, 2020) and Australia increasing from 821 to 'at least 1,245 workers and contractors' (IGEA, 2020). Kerr (2017: 99) notes that the 'overall increase in industry numbers masks the rate of turmoil in the industry and the numbers of company closures and start-ups.' Despite this, if near doubling of Kerr's figures holds across most of the industry, then we can probably estimate that the global games industry directly employs close to half a million people worldwide and be confident we are roughly in the right ballpark. As well as Kerr's (2017) earlier caveat about potential undercounting, and the fact that these figures do not include many parts of the world, this figure also discounts the many related and indirect jobs that the games industry supports. The Entertainment Software Association of America claims that the games industry in that country alone supports over 428,000 jobs (ESA, 2021), suggesting that for every direct employee as many as 5 other support workers are employed. Likewise, as Keogh (2021) has shown in Australia, there exists a substantial number of solo-developers, student developers, and others on the periphery of the game industry that make—or are aspiring to make—games which might also be included in estimates. At one point in 2016 the game development system Unity claiming to have over 4 million registered developers, many of them students perhaps, but many more solo and small team developers who are not directly employed, but who are nevertheless part of the industry (Unity, 2016). Not all of these will have the same emissions intensity profile as employees of Triple-A game developers, naturally, but they are unlikely to make *zero contribution* to the overall figure, and should therefore be counted in the industry's footprint. Further research will be needed to more accurately establish whether the emission intensity of support roles, student developers, and solo developers working from home offices match the figures outlined in this chapter.

We're left then with either a conservative figure of around half a million directly employed game developers worldwide (most likely an undercount) and perhaps as many as 3 or even 4+ million direct and indirect workers, such as those on contracts, freelance, self-employed and other workers peripheral to the industry (and which might risk stretching the applicability of our identified emissions profile). Nevertheless, I think we can be confident that at least half a million developers work in the games industry in or close enough to a traditional game development context, and for whom our estimated games industry *per employee per annum* figure is applicable, and perhaps even substantially more than that.

If we assume that the larger companies on the scale (Ubisoft, EA, Nintendo, etc.) are fairly representative of the typical core games industry worker, and that our sample of home and office based workers from Australia (which is generally an emissions intense country anyway) is likely to be an outlier, with solo developers figures representing combined personal and work energy usage, we might figure that a reasonable estimate of emissions is somewhere in the range of **1–5 tonnes** of $CO_2$ equivalent per employee per annum. To produce our estimated footprint we multiply this range with the number of employees in the global games industry. Starting with the conservative end of our estimate—this works out to be a range from **0.5 million** to **2.5 million tonnes** of $CO_2$ equivalent. This is the *bare minimum* emissions that the games industry is guaranteed to have produced and to be responsible for emitting into the atmosphere annually. If we tentatively, and at the risk of introducing inaccuracy, count the global games industry closer to the 3–4 million figure (or even more), that shifts our figures to a range of **3–15 million tonnes** of $CO_2$ equivalent emissions per annum.

This seems all within the realms of possibility, and far off other similar size industries. If this higher estimate is accurate, then that would put the game development industry, its workstations, server rooms, and everything else, at around the same emissions intensity as the total 2018 emissions for the European country of Slovenia, with a population of about 2 million people. If the games industry were its own country, it would be somewhere around the 130th most intense emitter in the world—an impressive if unwelcome feat.

This figure compares with other entertainment industries like film and television. The Sustainable Production Alliance released a report in March 2021 with figures for film and TV production emissions, with average emissions per feature for blockbuster "tent-pole" films (budgets over $70 m: 3370 tonnes of $CO_2$ equivalent) all the way down to small films (budgets less than $20 m: 391 tonnes of $CO_2$ equivalent) (Sustainable Production Alliance, 2021). Unfortunately, the report does not provide an overall sector emissions figure, so we can only make estimates based on similar guesswork to the above. According to the European Audiovisual Observatory (2019) there were 5823 films produced in 2018 in the leading film markets worldwide—if we took the "medium" sized film feature production (budget of $20–40 m: 769 tonnes $CO_2$) as an average then a rough estimate of the global film industry's carbon footprint would be somewhere around 4.4 million tonnes of $CO_2$ equivalent per annum. This would put global film emissions at the low end of our possible range of emissions from game production, and if the global distribution of films produced were closer to the "small" budget films this figure falls even further to only 2.3 million tonnes of $CO_2$, almost certainly lower than the global games industry.

It is also my strong suspicion that our lower-bound figure of 0.5 m tonnes of $CO_2$ for games development represents a serious under-estimation, given that there are many, many game companies not included in this chapter's discussion, most of whom do not appear to have begun to report emissions let alone take active steps to reduce them. If it is safe to assume that those companies which have begun to report their emissions are the ones which are making the most progress on, then we might lean towards the middle or even upper end of the emissions intensity range. It is also important to remember that this figure also *only* includes the direct Scope 1 and 2 emissions from game developers and omits large parts of what makes modern game development possible, like third party data cen-ters, flights for an increasingly global industry (fuel and flights being counted in the film industry figures), hardware manufacturing, and the eventual end players themselves which contribute substantially to the industry's overall footprint. As the Rovio CSR reported noted earlier, air travel alone was triple the per-employee carbon emission figure. Likewise, Andrae and Edler (2015: 133) estimate, quite astonishingly, that alone 'data centers will use around 3–13% of global electricity in 2030'—growth which undoubtedly games have contributed to significantly. The inclusion of these other emissions sources would blow out the industry's carbon footprint substantially, sending it rocketing past our film industry esti-mates. Given all this, I feel confident claiming that the global games indus-try is almost unequivocally more emissions intense than the global film industry, and perhaps even by a substantial margin.

According to the 2020 Global Carbon budget, the world's total emis-sions in 2020 were 34 billion tonnes of $CO_2$ equivalent. At the highest end of our emissions estimates, game development alone would therefore be responsible for about 0.04% of the entire world's emissions, despite employing perhaps a few million people. It is tempting to see this small fraction as inconsequential, however it should be sobering, a wake-up call to an industry that has by-and-large been complacent about its global impact in the face of decades of warnings about rising greenhouse gas emissions and other environmental issues. Clearly a huge challenge lies ahead of the games industry—whether it is 0.5 or 15 million tonnes of $CO^2$, this is a lot of it to deal with, representing an offsetting cost of €25–750 million at today's carbon price of €50 a ton. Preventing those emissions in the first place is going to be much, much cheaper than trying to offset them after the fact. The urgency of a transition to a carbon neutral games industry is now clear, and the task before us is now about identifying potential pathways to decarbonization.

## CONCLUSION

This chapter has attempted to describe the source and scale of emissions associated with making the modern digital game, gathering data for a range of different games and developers, making games on platforms ranging from consoles, to PC and mobile devices. We have observed a range of figures and emissions targets, drawn from the responses to the survey of energy use and intensity in game development, combined with the increasing trend to report annual emissions footprints in CSR reports. From this we see that the emissions intensity of the digital games sector is substantial. A whole-of-industry figure remains an estimate at best, hopefully increasing in accuracy as greater transparency and disclosure is adopted more widely. What we have seen thus far is an emissions range, with both low and high per-capita per-annum intensities that offer a benchmark for other developers to measure against and up their game. We saw that many solo developers appear to have a high per-employee energy use or unknown consumption data, with energy use and emissions intensity figures inseparable from personal, domestic energy use. Small scale game development thus presents a unique challenge to benchmarking and accounting as it happens in bedrooms and shared spaces that are not dedicated office environments. Given the international games industry is now concentrated at both extremes of the scale, as seen in the State of the Industry surveys, (GDC, 2020) plus the rapid increase in work-from-home, the challenge of accurately measuring and acting on the emissions of this part of the industry may require specific, targeted approaches. At the other end of the spectrum, however, we see large publishers and multinational studios employing hundreds or even thousands of workers, some already achieving substantial reductions in emissions just by purchasing renewable power. A few of these companies are taking steps towards carbon neutrality, or even carbon negativity, and the power of large corporations in the games industry to influence investment in renewables should not be overlooked. In between these two ends of the game industry we still have a wide variety of studio profiles and emissions intensities, typically reliant on local energy grids for power.

Amongst the developers who responded to the survey there appeared strong support for a wider trend of disclosing game development emissions, only 2 out of the 30 respondents said they would not support this. Both were individual developers—one stated their reason as related to the privacy of personal data, another for it not being 'worth the effort.' Space Ape Games' Nicholas Walker notes his company's support for wider disclosure of emissions in the games industry which also supports the view that it become be an uncontroversial practice. He states that:

> I support them and we have broad support internally. We have ten years to substantially change the way we do business to reduce carbon emissions. Lots of that burden lies with government and much larger suppliers than we can impact, but we feel a need to do what we can and encourage others in the industry to follow suit.

Government legislation and a whole of system approach to decarbonizing energy grids around the world are also going to be necessary—however there are encouraging signs, and convincing technical and theoretical work being done to demonstrate that 100% renewable energy systems are possible. An increasing focus on efficiency will be part of this picture as well, and is already paying dividends, and this can also be driven by developers themselves and the decisions they make around the games they design—a topic we will return to in the second half of Chap. 6.

More work is still needed to fully account for emissions from air travel, to measure exactly how much this Scope 3 emissions source increases game development's carbon footprint and to develop approaches for mitigating the need for them. If our goal is achieving a carbon neutral games industry, then game development must be the first place we aim to achieve this goal, being both the emissions that the business of making games is most directly responsible for, and most it is immediately able to address. As I argued in the preceding two chapters, there can be no truly green games that are made with and powered by fossil fuels, and the challenge presented to the industry by these numbers and figures above is a substantial barrier to the truly ecological game. In the next chapter, I turn to the emissions from distributing games, both through sending discs around the world and via digital downloads.

REFERENCES

Andrae, A. S. G., & Edler, T. (2015). On global electricity usage of communication technology: Trends to 2030. *Challenges, 6*(1), 117–157.

Aslan, J. (2020). *Climate change implications of gaming products and services.* Ph.D. Dissertation, University of Surrey.

Compensate. (2021). *Reforming the voluntary carbon market: How to solve current market issues and unleash the sustainable potential.* https://www.compensate.com/reforming-the-voluntary-carbon-market

Cook, D. (2021, February) Video game software publishing in the US: Industry report 51121E. *IBIS World.* https://www.ibisworld.com/united-states/market-research-reports/video-game-software-publishing-industry/

Dept of the Environment and Energy. (2019, August). *National greenhouse accounts factors.* Australian Government Department of the Environment and Energy.

DEV. (2020). Libro blanco del desarrollo espanol de videojuegos. *Desarrollo español de videojuegos.* https://www.dev.org.es/publicaciones/libro-blanco-dev-2020

EA DICE. (n.d.). *Sustainability report.* https://static1.squarespace.com/static/5d9eed71525d4d25aaeeb997/t/5f62142864c11d7afeb94267/1600263209374/EA+Digital+Illusions+CE+AB+-+Sustainability+Report.pdf

Electronic Arts. (2019). *Electronic arts Inc.* Fiscal Year 2019 Proxy Statement and Annual Report. https://ir.ea.com/financial-information/annual-reports-and-proxy-information/default.aspx

Electronic Arts. (2020). *Impact Report.* https://www.ea.com/news/sharing-our-first-electronic-arts-impact-report

Erol, O., & Çirak, N. S. (2020). What are the factors that affect the motivation of digital gamers? *Participatory Educational Research, 7*(1), 184–200.

ESA. Videogame impact map. *The Entertainment Software Association.* https://www.theesa.com/video-game-impact-map/. Accessed 20 Aug 2021.

European Audiovisual Observatory. (2019, May 4). Leading film markets worldwide from 2007 to 2018, by number of films produced. Chart. *Statista.* https://www.statista.com/statistics/252727/leading-film-markets-worldwide-by-number-of-films-produced/. Accessed 20 Aug 2021.

GDC. (2020). *State of the game industry.* https://reg.gdconf.com/gdc-state-of-game-industry-2020

Golding, D., & Van Deventer, L. (2016). *Game changers.* Simon and Schuster.

Gunningham, N., Kagan, R. A., & Thornton, D. (2004). Social license and environmental protection: Why businesses go beyond compliance. *Law & Social Inquiry, 29*(2), 307–341.

IEA. (2020). *World energy outlook 2020.* International Energy Agency. https://www.iea.org/reports/world-energy-outlook-2020

IGEA. (2020, January 28). Australian video game development: An industry snapshot FY2019–20. *Interactive Games and Entertainment Association.* https://igea.net/2021/01/australian-game-development-industry-counts-185-million-in-revenue/

Keogh, B. (2021, January). The cultural field of video game production in Australia. *Games and Culture, 16*(1), 116–135. https://doi.org/10.1177/1555412019873746

Kerr, A. (2017). *Global games: Production, circulation and policy in the networked era.* Routledge.

Lee, A., Chen, O., & Tsai, J. (2013, September 4). Flextronics lands 90% of Xbox one orders, leaving Foxconn the rest. *DigiTimes.* https://www.digitimes.com/news/a20130904PD219.html

Lee, A., & Tsai, J. (2017, October 5). Nintendo ups switch supply to 2 million a month. *DigiTimes.* https://www.digitimes.com/news/a20171005PD202. html. Accessed 30 Mar 2021.

Microsoft. (2017). *Data factsheet: Environmental indicators.* http://download. microsoft.com/download/0/0/6/00604579-134B-4D0E-97C3-D525DFB7890A/Microsoft_2017_Environmental_Data_Factsheet.pdf

Microsoft. (2020a). *2020 Microsoft corporate social responsibility report.* https://aka.ms/2020CSR_Report

Microsoft. (2020b). *Microsoft devices sustainability report FY20.* https://aka.ms/devicessustainability

Microsoft. (2020c). *Microsoft sustainability calculator.* https://www.microsoft.com/en-us/sustainability/sustainability-guide/sustainability-calculator. Accessed 30 Mar 2021.

Microsoft. (2020d). *Environmental sustainability report.* https://www.microsoft.com/en-us/corporate-responsibility/sustainability/report. Accessed 10 July 2021.

Mishkin, S. (2013, Nov 13). Foxconn profits beat expectations. *Financial Times.* http://www.ft.com/cms/s/0/ea1cdd4c-4c70-11e3-958f-00144feabdc0. html. Accessed via archive.org 30 Mar 2021.

Nintendo. (2020, July). *CSR Report 2020.* www.nintendo.co.jp/csr/en. Accessed 30 Mar 2021.

Phillips, A. (2020). *Gamer trouble: Feminist confrontations in digital culture.* NYU Press.

Provasnek, A. K., Sentic, A., & Schmid, E. (2017). Integrating eco-innovations and stakeholder engagement for sustainable development and a social license to operate. *Corporate Social Responsibility and Environmental Management, 24*(3), 173–185.

Rovio Entertainment Corporation. (2020, March 6). *Corporate responsibility report 2019.* https://investors.rovio.com/sites/rovio-ir-v2/files/06-03-2020/eng/Rovio-CSR-report-2019.pdf. Accessed 30 Mar 2021.

Ruberg, B. (2020). *The queer games avant-Garde: How LGBTQ game makers are reimagining the medium of video games*. Duke University Press.

Ryan, J. (2020, August 31). No, Microsoft has not made 825,000 carbon-neutral Xbox Series X consoles. *CNet*. https://www.cnet.com/news/no-microsoft-has-not-made-825000-carbon-neutral-xbox-series-x-consoles/

Shaw, A. (2015). *Gaming at the edge: Sexuality and gender at the margins of gamer culture*. University of Minnesota Press.

Sony. (2020). *Sustainability report 2020*. https://www.sony.net/SonyInfo/csr/library/reports/SustainabilityReport2020_E.pdf. Accessed 30 Mar 2021.

Space Ape Games. (2020). *We've gone carbon neutral: Here's why we did it, and how you can too*. https://spaceapegames.com/green. Accessed 30 Mar 2020.

Sustainability Victoria. (2020). *Calculate appliance running costs*. https://www.sustainability.vic.gov.au/You-and-your-home/Save-energy/Appliances/Calculate-appliance-running-costs. Accessed 2 Mar 2020.

Sustainable Production Alliance. (2021, March). Carbon emissions of film and television production. *Sustainable Production Alliance*. https://www.green-productionguide.com/in-action/

Thomas, C., Rolls, J., & Tennant, T. (2000). *The GHG indicator: UNEP guidelines for calculating greenhouse gas emissions for businesses and non-commercial organisations*. UNEP.

Ubisoft. (2020). *Universal registration document and annual financial report*. https://www.ubisoft.com/en-us/company/about-us/investors. Accessed 30 Mar 2021.

UKIE. (2020). *Think global, create local – The regional economic impact of the UK games industry*. https://ukie.org.uk/regional-economic-report

Unity. (2016, July 30). The leading global game industry software. *Unity*. Internet Archive saved version. https://web.archive.org/web/20160730021122/http://unity3d.com:80/public-relations. Accessed 20 Aug 2021.

WBCSD, WRI. (2004). *The greenhouse gas protocol. A corporate accounting and reporting standard* (Rev. ed.). Conches-Geneva.

Westwood, D., & Griffiths, M. D. (2010). The role of structural characteristics in video-game play motivation: A Q-methodology study. *Cyberpsychology, Behavior and Social Networking, 13*(5), 581–585.

GAMES

Space Ape Games. (2015). *Rival Kingdoms*. Space Ape Games.

Space Ape Games. (2016). *Transformers: Earth wars*. Hasbro, Space Ape Games, Backflip Studios.

# CHAPTER 5

# The Carbon Footprint of Games Distribution

Once a game is made it still has to get into the hands of players—until quite recently, this almost always meant a game had to be sold in some sort of physical form, being first produced in a factory somewhere and then shipped to wherever games are sold. A cartridge, a disc, some more or less stable digital storage media—each has to be made from a material substrate (magnetic tapes, ROM circuits, polycarbonate discs with a thin layer of reflective foil) and combined with inks, paper, printed cardboard cartons, plastic cases and shrink-wrapping. Physical media has been critical to the history of games distribution, and even now forms a substantial part of game sales—and it's likely they will continue to for some time to come, important for a host of reasons, not least of all the second-hand games market. Increasingly, however, the sale and distribution of games happens online through a digital distribution platform like Steam, or the PlayStation, Xbox or Nintendo stores. These platforms are enabled by the physical infrastructure of one of a handful of companies that maintain content delivery networks (CDN's). These CDN's in turn maintain relationships with the companies that run and operate physical data centers and other digital infrastructure all around the world, providing the hardware layer and software tools that Steam and other customers tap into in order to serve up low latency, high-bandwidth streams of digital data to customers all around the world. CDNs push the physical location of the data and servers that users' access 'closer' to the end user, and allow a game made

B. J. Abraham, *Digital Games After Climate Change*, Palgrave Studies in Media and Environmental Communication, https://doi.org/10.1007/978-3-030-91705-0_5

in, say, Seattle to be downloaded at lightning speeds on the other side of the world. This is still a relatively new development, but is rapidly becoming the dominant method of game delivery.

Both digital and physical distribution methods have carbon emissions associated with them, and the goal here in this chapter is not so much to say which is necessarily 'better' exactly, but rather to evaluate the challenges that the industry faces in achieving carbon neutrality for both, and what that suggests for the near-to-medium term of the industry. The chapter has three sections. The first investigates the physical distribution systems and emissions involved in shipping games from manufacturing plants to the different parts of the world they are sold in, using the example of the PS4 disc distribution channels circa 2020. A number of researchers have provided life-cycle assessments (LCAs) of disc distribution, and in particular, attempted to answer questions around which method of distribution (physical or digital) is more or less carbon intensive (Mayers et al., 2015; Aslan, 2020) arriving at different answers based on different assumptions and prospective scenarios.

These existing studies have largely focused on the emissions of Europe and the UK, already close to the location of disc manufacturing however. The second section of the chapter undertakes a desk analysis of the potential emissions embodied in the distribution of discs to one country—Australia. This case study illustrates the emissions entailed in distributing games to the extreme ends of the world. These emissions levels may not be apparent based on the earlier studies, which might not be representative for disc distribution to the entire world. Further validation of these (entirely hypothetical) desk estimates is still required, preferably via full LCA methodology, in order to achieve an accurate picture of the emissions of this method of distribution for countries further away from the close manufacturing and logistical links of Europe. Undertaking an LCA of a particular *game* is also a challenge that remains beyond the scope of this book. The data collated here may however assist those who wish to do so.

The third and final section of the chapter discusses the carbon footprint of digital distribution, once again drawing on existing research and modelling, the details of which produce a complex picture depending on a variety of variables and assumptions. What this section offers however is more of a broad discussion of the overall dynamics and trends involved. As illuminating as the existing work has been, what should remain at the forefront of our analysis is where the greatest potential for emissions *reductions* lies as this is more important for climate advocates and achieving our goal of a carbon neutral games industry. To put it another way: what is more important is how easily, and how quickly, any given method of distribution could *become* carbon neutral, more than whether they necessarily are right now. This deeper structural question regarding the choices to be made in distribution is absent from current research which (for perfectly sensible reasons) is thus far confined to looking at the emissions associated with a given analysis period. The focus on the present, while a necessary first step in achieving an understanding of the *current* state of emissions, ends up discounting the changes that are already under way, and those which will become necessary in the very near future, many of which are only now becoming clear as renewable power generation technology matures. The current energy intensity of data centers, data transmission infrastructure, as well as the energy used by gaming devices in downloading from the internet, are examined in this section, drawing particularly on the pioneering work of Joshua Aslan (2020), who has undertaken detailed modelling of a variety of scenarios, producing detailed and robust estimates.

The intractability of physical media's transport emissions, coupled with the scale of digital infrastructure energy use and emissions revealed by Aslan's (2020) work, as well as other macro-trend energy modelers like Andrae and Edler's (2015) points to the utmost urgency of decarbonizing energy systems worldwide and in transitioning to renewable power generation. The chapter ends on a hopeful note, with a brief discussion of the evidence for the technical and economic feasibility of 100% renewable national energy grids that might get us there.

## THE EMISSIONS OF SENDING DISCS AROUND THE WORLD

In the earliest periods of the game industry games were first shared and disseminated as source code, usually in a book or magazine, with the end user having to type in the code themselves (Donovan, 2010; Kirkpatrick, 2012). As gaming became more accessible, moved out of the arcade, and into the home with the first consoles, the sale of games moved to storage media with a copy of the game on it—from cassette tape, to floppy disk, to CD-ROM. Games being sold in retail stores paved the way for the modern commercial games industry, coming to require exclusive access to distribution networks only afforded by a publisher. When most people think of games, home consoles and the proprietary hardware and software ecosystems they represent probably comes to mind. According to historical accounts, like Donovan's (2010) *Replay: The History of Videogames,* the modern certification system that gate-kept these platforms began in the 80s starting with the Nintendo Entertainment System. Reputedly developed in response to a "flood" of cheap and low-quality games that was said to have contributed to the great crash of '83, these certification systems meant that for the longest time making and selling games on consoles was a fairly specialized, capital intensive process (Donovan, 2010; Johns, 2006).

Making games in this way meant negotiating with the platform owners for permission to even develop games for their console, and conforming to their particular standards. This further entailed manufacturing orders, arrangements with suppliers, packing, freight and other logistics issues involved with getting games into shops and player's hands. The manufacture of large volumes of cartridges or discs has almost always been prohibitive to developers without a large and experienced publisher backing them, or their own extremely deep pockets (Johns, 2006). As physical objects these commodities have to be made somewhere, and in the current period of intense globalization that often occurs far from wherever the end consumer is in the world, inevitably entailing carbon emissions from transport. In the two main studies of *PlayStation 3* and *PlayStation 4* game distribution we consider here, the location of manufacturing occurs in Europe, not too far from their end consumer. But for game players in other parts of the world—East Asia, Australia, even perhaps North and South America—these discs can end up crossing large distances to reach the game purchasing public.

The two existing attempts to measure the emissions of different game distribution methods were conducted by Mayers et al. (2015) with calculation based on 2010 figures, and by Aslan (2020) with figures based on data leading up to 2020. Both studies were only possible through the cooperation of Sony and access to distribution centers, and present a unique insight into this part of the games industry. A truly landmark paper, the earliest of the two by Mayers et al. (2015) undertakes a detailed analysis of the carbon footprint of physical and digital distribution methods for *PlayStation 3* games distributed within the United Kingdom in 2010. The authors provide one of the few detailed descriptions of the manufacturing process, emissions intensity, and overall lifecycle of a typical Blu-Ray disc. It's worth quoting at length for understanding precisely how involved this process is, and the huge array of materials and labour marshalled:

Each BD is composed of a combination of polycarbonate, silver, and a protective resin. Mastering involves the transformation and projection of an encrypted and certified digital data file onto a silicon wafer. The BD mastering process happens only once per new game or film, from which thousands of replicated copies are made. The replicated discs are then transferred mechanically to the printing line for disc artwork and then transferred to the assembly packaging line. After the discs have been placed into a polypropylene molded box case with an inlay tray and a paper instruction booklet, they are packed into cardboard master cartons, stacked on wooden pallets, and secured with a polypropylene film wrap. Discs are then distributed by truck and shipped to a central warehouse in Northampton, UK and subsequently to retailers' warehouses ready for distribution to outlets and sold to consumers. Subsequent to use, domestic recycling options do not exist for BDs, and so at EOL they are collected and sent to either landfill or incineration. (Mayers et al., 2015: 2)

Ten years on and it is likely that some small changes have occurred, but this account still remains an important description of the production of physical copies of PS3, and (and now PS4 and PS5) games sold around the world today. Mayers et al. (2015: 5) note in their study that at the time 'in 2010, 95% of games were distributed by disc'—a figure that has shifted greatly in the decade since, with digital sales approaching parity in 2020, and in some cases even surpassing, physical game sales (Ahmad, 2020). In a sure sign of the direction things are headed, the launch of the PlayStation 5 console saw two versions of the device go on sale—one *without* an optic disc drive at all, only able to download games digitally.

The goal of Mayers et al.'s (2015: 2) study was 'to calculate the carbon equivalent emissions of the life cycle of PS3 games and also determine whether downloading data had lower impact than distribution by BD in 2010.' The study attempts to account for the 'carbon equivalent emissions arising from the raw materials production, manufacture, distribution, retail, use, and disposal of new PS3 game BDs and files produced for the UK' (Mayers et al., 2015: 3). The list of measured $CO^2$ equivalent emissions for Blu-ray manufactured and distributed discs in 2010 is tallied, leading to a total of 1.2 kg of $CO^2$ equivalent emissions per disc in 2010. This seems a relatively modest amount, however we must remember that this is just one PS3 disc, hundreds of millions of which were sold. According to VGChartz (2021), over the lifetime of the PS3 console there were 346.1 million games sold in the European region over the lifespan of the console (this was the closest matching data I could find, and the most comparable to Mayers et al.'s (2015) focus on the UK). Some, though perhaps not many, of these sales will have been digital, however if these figures are close enough to apply to European PS3 game sales, then making and distributing all the games over the lifetime of the PS3 came with some significant carbon emissions—perhaps as much as 415,000 tonnes of $CO_2$ emissions. Distributing all of Europe's PS3 discs over the lifetime of the console therefore emitted around the same as a small English city of around 75,000 people (like Guildford just outside of London) emitted in a whole year.

In a similar study, updating these figures for the decade since, and encompassing a greater range of other factors and variables, Joshua Aslan (2020) undertook a lifecycle analysis of PS4 games in a variety of scenarios for his doctoral research project. Aslan's (2020) work represents a substantial development from Mayers et al.'s (2015) original scope, providing a range of different scenarios that helpfully illustrate the dynamics involved in different distribution methods (including the slightly more relevant prospect of cloud streaming services—though I omit these here as not particularly relevant consideration at present). Aslan's (2020: 193) figures for disc-based distribution are somewhat higher than the previous study and which he tallies up as follows: 0.27 kg for Disc Production, 0.00 for Disc Distribution, 2.04 for Disc Retail, and 0.18 for Disc Disposal, for a total of 2.49 kg of $CO_2$ equivalent emissions—a doubling of Mayers et al.'s (2015) estimate from a decade earlier. Partially these differences may be explained by the different system boundaries between the two papers, differences in calculation methodology, and so on, however this is beyond the scope of our discussion. However what is important to underscore is at the very least these figures are *not decreasing*. This could not be further from the case when it comes to digital distribution emissions figures, as we shall see in the third section of the chapter, when we look at both studies figures for digital distribution and the reasons for the change from 2010 to 2020.

But before we get there it is important to acknowledge a couple of important factors involved in both these studies calculations—firstly, the Sony Blu-ray disc pressing plant in Austria is significantly closer to Europe and the UK than other parts of the world that it ships discs to (like Australia), resulting in lower estimates for transport emissions than will be the case elsewhere. Secondly, these emissions from manufacturing and transport distribution are the sort of emissions that, while possible to 'offset' with paid credits for reductions from elsewhere, it is not immediately evident that they can be made completely emissions free, which would be better than emitting and offsetting. These factors becomes even more pronounced when we consider the logistics of shipping games to the other side of the planet, as in the case of shipping games to Australia, which I now turn to as a case study illustrative of the emissions entailed by shipping to regions further from manufacturing hubs.

## A Case Study of Australia: Estimating Disc-Based Distribution of PS4 Games

We know from Aslan's (2020: 160) research that all *PlayStation 4* Blu-Ray discs in the world are manufactured by SONY DADC Europe's factory in Thalgau, Austria and that encouragingly the plant is now powered by 100% renewable energy (Sony CSR, 2020). This appears to be a relatively recent change, with Sony previously manufacturing discs (now just assembling them) in an Australian factory located in Huntingwood, an industrial suburb in the west of Sydney, until it closed the manufacturing aspect of the site around 2018. Prior to this, PS4 games sold in Australia included small print on the back of the case with a fine-print addendum noting they had been 'manufactured in Australia'—for instance, my PS4 copy of *Dark Souls 3*, (2016) has this designation. More recent PS4 games like the *Final Fantasy VII Remake*, (2020) however, change this text to state 'manufactured in Austria. Assembled in Australia'. It is entirely possible that this centralization change might even produce a reduction in overall emissions (at least from a strict GHG accounting perspective), particularly given the 100% renewable powered Thalgau plant. The centralization of manufacturing, however, does allow us to know the distance that all recent PS4 discs (and now presumably PS5 discs) *unavoidably must travel* to reach to Australia, entailing a certain level of emissions intensity which we can attempt to estimate.

As with any logistical operation involving the transport of consumer goods, there are largely four shipping options available to Sony to get Blue-Ray discs from Thalgau, Austria to customers in Australia: road, rail, sea, and air. Given the location of the factory in Thalgau, road-based freight will be involved for the first and last portion of the distance at both ends, however Australia is also on the other side of the planet, and separated by oceans. Rail shipping requires extensive existing infrastructure, does not cross oceans, and is unlikely to feature in this scenario.

Alternatively, the city of Vienna, located on the river Danube, is only 300kms away from Sony's Thalgau plant via the A1 Autobahn, and has a large container shipping terminal for transfer to one of the many container ships that travel up and down the river. This shipping route leads out to the Black Sea, and from there ships may pass through the strait of the Bosporus and the Dardanelles out to the Mediterranean Sea, to then (in the most direct route over sea to Australia) pass through the Suez, the Red Sea and the Gulf of Aden, and eventually reaching the Indian Ocean and Australia. This (hypothetical) journey, however, would take several weeks, with one shipping estimate from the port of Gemlick in Turkey to Brisbane in Australia suggesting an estimated travel time of 48 days, though estimates vary. If shipped via seagoing vessel, we are faced with the tricky task of estimating maritime shipping emissions, which can be difficult to measure without also considering factors like stops at different ports, speed of travel, engine efficiency, use of different fuels, and so on (van der Loeff et al., 2018). As van der Loeff et al. (2018: 896) also note, 'maritime transport offers by far the most energy efficient mode of long-distance mass cargo transportation.' It is primarily the capacity to carry large volumes of heavy cargo, however, that makes this form of shipping attractive and even necessary for some goods, however as we shall see in a moment, these considerations may not match those of the fairly light and quite compact PlayStation 4 discs.

Conveniently, Vienna international airport is also Austria's largest for air freight, (Centre for Aviation, 2021) and is our final method of getting discs to Australia. At least three air freight companies operate out of the airport. With only a few cargo planes capable of a direct flight from Europe to Australia, at least one stop over will almost certainly be required for refueling, usually somewhere in South East Asia—Bangkok or perhaps Singapore. Google Earth measures the straight-line distance from Vienna to Bangkok at about 8400 km, and another 7500 km from Bangkok to Sydney for a total distance of about 16,000 km travelled. This seems a plausible and conservative estimate given other factors that might come into play, like detours to avoid weather, head and tailwinds, the specifics of flight paths, and so on which would impact final distance travelled. This figure also does not change substantially whether the flight goes via Singapore's Changi Airport instead. In both cases, approximately 16,000 km's are traveled by these discs as air cargo.

Arrival in Sydney is not the end of the story however, as it then needs to be unloaded and freighted from the terminal to the same Sony factory for assembly, packing. This is another 50 km trip—almost certainly via truck—where the discs will be unpacked and put through an assembly line where these individual discs are placed into the iconic blue plastic PS4 cases (we will ignore the manufacturing and shipping of plastic cases for our scenario), along with the printed booklets and jacket material that displays on the outside of the case. Each case is then individually shrink-wrapped in plastic and placed in cartons or boxes for transport to various warehouses for distribution to the retail shops it will eventually be sold in. It can then be sold to a customer who will part with their money, and take the game home, completing a massive journey across the globe.

Before attempting to estimate the emissions that are entailed by all this travel I want to reflect on what this already shows. By highlighting even the most general outline of these logics chain involved in shipping game discs around the planet, we see already how important a whole-of-system perspective is for our task, and just how much escapes the drawing of a 'system boundary' for accounting. The expansive nature of the climate challenge itself demands that we look at the picture in the most systematic way possible, or risk shifting the responsibility for emissions simply to someone else—this could not be further from the idea of thinking eco-logically. This sort of analysis means including and accounting for entire supply chains: like where the raw materials have come from, what emissions are embodied in them before they even arrive, and what sorts of other harms lie just outside the boundary of direct corporate responsibility. As we saw in the previous chapter, multinational corporations like Microsoft and Sony can have an influence on their suppliers, which are often companies that are not visible to the public, and as such may face less direct pressure to decarbonize from a climate conscious public. Many of these companies are not household names, and finding ways to pressure them to clean up their act might have to be done through the large corporate purchasers. Whatever the strategy, and corporations will need to look to their suppliers, as well as at the up- and downstream emissions they are creating, whether through demand or design, for both plastics and power.

What sort of carbon emissions might be embodied in Australian PlayStation games after travelling this far? Having a sense of this hypothetical (though very real) journey that each and every PS4 Blu-ray manu-factured disc that reaches Australia has been on, at least since manufacturing shifted in 2018, what sort of emissions are involved? We can tally up the distances traveled in different modes of transport and estimate emissions for each.

As a plausible estimate of the weight of our cargo, based on the size of a disc (20 cm in diameter, 1 mm in thickness) it seems possible to fit around 40,000 discs on a pallet (dimensions: 101 cm × 122 cm) and about 20 pallets to a truck—perhaps 800,000 discs in all, carrying no other cargo (helpfully, for simplicity's sake, keeping our scenario to just one truck load of discs). Each disc weighs on average between 15 and 16 grams (based on my own measurements of my PS4 game disc library using a scale accurate to 0.1 grams). Assuming 15.5 g per disc, this adds up to approximately 13 tonnes worth of PS4 discs.

The European Environmental Agency's emissions factor figures for heavy diesel trucks suggest emissions of 0.486 grams of $CO^2$ per km. The first leg of the freighting process from the Thalgau plant to the Vienna international airport, for one truck trip, adds up to about 145.8 grams of $CO^2$ equivalent emissions. This is a small figure; however the journey has only just begun. How the discs cross the oceans, however, makes the much larger difference to overall emissions.

As suggested earlier, our 13 tonnes of PlayStation discs *could* be shipped via sea going cargo vessel, and if so the emissions would be substantially lower than by air. According to a 2009 study by the International Maritime Organization shipping via 'very large container vessel' can be as low as 3.0 grams per tonne per km (World Shipping Council and the International Maritime Organization, 2020). Given 13 tonnes of discs, and a journey of almost 17,000 km (measuring from Constanta, Romania at the mouth of the Danube, to Sydney, Australia) according to the Sea Distances shipping calculator (Sea-Distances.org, 2021) this represents a journey of almost 40 days at 10 knots (about 18 km/h). Assuming no interruptions to this journey, and no detours or time spent in ports in North Africa, the Middle East, or South East Asia along the way, this results in about 663 kg of $CO^2$ for our 13 tonne shipment (or about 0.82 grams per disc). While this is again a fairly negligible figure on a per-disc basis, these do add up given the scale of the games industry and its impressive sales figures. Furthermore, two examples strongly suggest to me that sea-borne shipping methods are not the preferred for shipping most PlayStation discs to Australia. The time from when a game goes 'gold' and is functionally 'finished' and ready for manufacturing, and the date on which it goes on sale in stores is simply too short to accommodate a lengthy oceangoing trip. The first example is from Sony Santa Monica studios creative director Cory Barlog (2018) who made a post on the PlayStation blog on March 22nd, 2018 noting that *God of War* had just gone gold, and was ready to be manufactured.

The game was on sale in Australia by the 20th of April, a mere 30 days later. Even if the Sony DADC plant immediately prioritized the discs destined for Australia at launch and had them ready to ship on the very same day Barlog made his blog post, the transit time to Australia by sea would blow-out the release window, taking at minimum 38 days from the Black Sea to Sydney, and assuming no stops on the way. It also does not take into account the time for assembly and packing, and how long it might take to complete nation-wide distribution of discs to stores around a country as big as Australia. Another example follows a similar pattern, with the *Cyberpunk 2077* twitter account tweeting about going gold on October 5th 2020, the game being on sale in store by Nov 19th (CyberpunkGame, 2020)—while this is a slightly longer lead time of 42 days, it is still implausibly quick for sea borne shipping given the other factors mentioned. Given these short lead-times from going gold to being on sale in stores, I am confident that PS4 discs (at least for their very first shipments of new releases, if nothing else) do not arrive in Australia via ocean freight, which leaves us with our final emissions intense option—air cargo freight.

The transport chapter (Chap. 8) of the 2014 IPCC mitigation report provides a range of emissions associated with long haul air freight, such as may be the case on the trip from Vienna to Sydney, estimating a range from 350 grams of $CO_2$ per km travelled, to up to nearly 1 kg per km (IPCC, 2014). Fortunately, region specific data is available, with Howitt et al. (2011: 7041) calculating emissions factors for the fairly comparable case of New Zealand air freight at '0.82 kg $CO_2$ per tonne-kilometer (kg $CO_2$ per t-km) for short-haul journeys and 0.69 kg $CO_2$ per t-km for long-haul journeys, based on New Zealand fuel uplift data.' Assuming these figures are roughly equivalent to their Australian counterparts, and assuming the lower figure of long-haul flights applies (let's say 690 g is emitted per km), for our flight of about 16,000 km that results in about 143.5 tonnes of $CO_2$ equivalent emissions per flight. The local stretch from Sydney airport to factory assembly line entails another truck journey of about 50kms, perhaps only another 24 grams of $CO_2$ equivalent in total, and easily as much again (if not more) after it is assembled and trucked around the country (for simplicity's sake we will leave this last leg out as these discs are likely to be transported alongside other freight by this point, complicating the individual footprints).

To estimate the emissions on a per-disc basis we need to know how much of each mode of transport is utilized solely for the freight in question, both in terms of weight and in terms of space—this complicates things, so I shall try to keep it simple. The first leg of the journey via road freight we estimated emissions of 145.8 grams of $CO^2$, divided amongst nearly a million discs, which results in an almost negligible emissions per-disc, barely worth rating a mention—only cumulatively does this add up. The emissions per-disc stops being negligible however once we come to the air-freight leg. The 747-400F cargo plane can carry approximately 128 tonnes of cargo, however it only fits 30 pallets by volume (AirBridgeCargo, 2020), meaning that probably 2/3rds of our cargo capacity is devoted to discs, assuming that this entire shipment of 800,000 discs is going to Australia. Assuming that the aircraft emits around 143.52 tonnes of $CO^2$ over the 16,000 km journey, we arrive at a final figure for each disc's embodied emissions (assuming it consists of 2/3rds of the overall cargo) of about 0.179 kilograms of $CO^2$ per disc. Doesn't seem like a whole lot, admittedly, but this figure allows us to broadly estimate for the *entire* number of discs that have been shipped and sold in Australia in a particular year (assuming that all discs are freighted via air—a plausible supposition, given just-in-time manufacturing trends, but not guaranteed). Suddenly the car-bon footprint of distributing physical discs begins to look a lot more impactful, and less viable as part of a carbon-neutral games industry.

In the annual Entertainment and Media report (PWC, 2020), PricewaterhouseCoopers reported $365 m worth of physical console game sales in 2020—assuming an average sales price of $80 AUD (a very conservative estimate, given many games are sold for much less, increasing our estimate of how many discs are being flown into the country) we get a figure of at least 4.5 million console games sold in Australia in 2020. This aggregate figure will necessarily include discs and cartridges, includ-ing for consoles other than the PS4, each with their own different supply chains and journey to make raising or lowering these figures. But for the sake of simplicity, let us assume that these other discs and cartridges have similar $CO^2$ emissions associated with them. If this per-disc figure is repre-sentative then all the games shipped to and sold in Australia in 2020 may have an associated carbon footprint of around **800 tonnes** of $CO^2$ equiva-lent emissions. This is, in all likelihood, an underestimation given just how much is left out of this very simplified calculation. More research would be needed to validate this, but even so this figure does not include a great many other logistical components of the process, like shipping packed and shrink-wrapped cartons of games to stores around the country, or the emissions of customers going to and from the store to purchase the game. These would surely see this number climb even higher.

The analysis above is a mostly hypothetical desk exercise, an attempt at establishing rough carbon costs for shipping to a remote part of the world, and still needs verification with real-world measurement and preferably a methodical lifecycle assessment akin to what Mayers et al. (2015) and Aslan (2020) have completed for European and UK distribution. What's more important perhaps than an exact figure, however, is what these emissions point towards. If our goal is a carbon neutral games industry, then we need to identify credible pathways to bringing this number down. Though we might be tempted to make a change to mode of transport, like avoiding shipping via air, surely the better option long term is to stop shipping these products *all together* and replace them with digital downloads. This will be a contentious position, made more difficult by the fact that both Mayers et al. (2015) and Aslan (2020) identify specific scenarios where disc distribution has a *smaller* carbon footprint than digital downloads (as we shall see in the next section). However the devil is in the detail in the system boundaries, as these only consider the very interconnected region that is Europe and no other far flung territories. Furthermore, and perhaps more importantly, the assumptions that emissions factor assumptions that underpin both studies are rapidly changing as grid-scale renewables increase as a proportion of the world's energy system.

The analysis above should not be treated as gospel, the numbers are far from definitive but they are illustrative of the problems facing us in the task of decarbonization of games distribution. Even this partial account of what is involved in the distribution of discs from one side of the planet to the other makes staggeringly obvious that the physical distribution of games is almost certainly incompatible with our goal of a carbon neutral games industry. As hard as this may be for us to accept, particularly for enthusiasts and collectors and the companies that market "deluxe" editions of games with pre-order incentives, and so on.

Whatever the transition looks like, however, the burden to demonstrate that the distribution of physical game media can be done in a sustainable and carbon neutral way now rests squarely on those that desire to see it continue. For those wishing to prioritize a livable planet and keep dangerous global warming in check, we will need to prioritize and accelerate the trajectory towards full digital distribution. The conclusions here add further weight to the calls of others, like Moore (2009), Taffel (2012) and others who have raised the issue of increasing waste burdens (both plastic and electronic) generated by digital media and games, to which a greater portion of digital sales could help alleviate.

In the next section we look more closely at the emissions associated with digital distribution, which while currently higher in some scenarios, are primed to benefit enormously from the rapid transformation of energy generation systems to increasingly cheap and clean renewable power.

## THE EMISSIONS OF DIGITAL DISTRIBUTION

In simplest terms, digital distribution involves a website or online store, from which consumers purchase games. The files for the game are downloaded from a server and then stored on your local device until you delete or uninstall it. Very simple in principle, however in practice it involves a host of technologies, literal world-spanning infrastructure, dizzying arrays of different software and hardware configurations and standards, and multiple multinational companies—potentially across more than one regulatory jurisdiction. One of the first digital distribution platforms for games, the success of which almost single-handedly served as the proof of concept for digital game distribution, was Valve's Steam service. It's success, along with Apple's App Store for the iPhone, would pave the way for others—like Microsoft, Nintendo and Sony—with their own platforms. Valve's first-mover advantage and developments in the technology and infrastructure of Steam, has placed it at center of the PC game distribution market for well over a decade, accounting for as much as 75% of the PC game sales by 2013 (Edwards, 2013). Though some still prize the physical aspects of game collecting, (Toivonen & Sotamaa, 2010) it seems likely on current trends to be more and more about digital downloads.

The multiple appeals of a digital download service for games is now obvious. One no longer needs to leave the house to acquire a new game, plastic cases no longer clutter up one's house, and one has access to an ever-increasing selection of new games and a back-catalogue of older titles. What one gives up (with a few exceptions) however is the ability to resell, lend or gift games to others once finished with them, and there always remains a risk of losing access to one's account for one reason or another (theft or bad behaviour leading to a ban from the service, for instance). There are good reasons to be wary of the market power that platform holders posess over digital storefronts—problems that Srnicek (2017) describes as 'platform capitalism.' Indeed one brake on such a transition may well be consumer desires themselves, as Toivonen and Sotamaa (2010: 205) note in their study of gamers attitudes towards digital downloads, 'as long as the rights to refunds, resells and loaning are constricted the allure of digital copies will remain limited.' While the industry seems to have overcome some of the early resistance to the downsides of these tradeoffs, there remain ongoing and unresolved tensions that may play a part in determining what future is possible or acceptable for game consumers and the game industry's digitization as part of its carbon neutral plans. While it is my contention that a carbon neutral games industry will, by necessity, primarily involve digital downloads, there still remain challenges in ensuring this transition is made equitably, including the difficult task of reckoning with the power of platform holders. Daniel Joseph's (2018: 704) study of Valve's abortive attempt at monetizing the work of *Skyrim* modders concludes that digital distribution platforms (particularly Steam) are active sites of struggle between labour and capital, the event bringing 'to the forefront a series of social contradictions that had previously been latent in the practice of modding.' Similarly, the carbon neutral games industry must not come at the expense of our collective control over our hardware and software. It is not a just transition if it comes at the expense of developers either, beholden to paying rents to platform owners and subject to their whims as they gatekeep access to markets and devices. Gillespie's (2010) analysis of the political dimensions of 'platforms'—even the discursive construction of hardware and software as *platforms*—suggests these deserve much greater scrutiny. We cannot uphold climate justice by solving one problem be entrenching the same tendencies of unconstrained monopoly under capitalism.

In this part of the chapter I attempt to gain a better picture of the energy and emissions involved in digitally distributing a game, and what sort of emissions are entailed in that process. As the two previous studies by Mayers et al. (2015) and Aslan (2020) show, this is a complicated task. Similar to the previous discussion, the end results of calculations for digital distribution once again depend on a number of assumptions that can change the outcome quite dramatically depending on where in the world a game is being downloaded, making accurate accounting and comparisons across the globe complicated.

Mayers et al. (2015) offers the following figures for digital download emissions in 2010: 'for an average 8.80-GB game... 21.9 to 27.5 kg.' (Mayers et al., 2015: 12) For the purposes of their study they chose to include gameplay energy in their analysis, which I would rather address separately as this is a separate variable (and which I return to in the first half of Chap. 6). Almost all modern consoles now install data and read from local storage media even if a disc is present, rendering the operational differences minimal: Mayers et al.'s (2015) analysis notes only a tiny difference of 10–15 watts in the PS3's energy usage for disc-based gaming for instance. The authors acknowledge that the difference amounts to basically nothing, stating that 'gameplay (use phase) accounted for 19.5 kg $CO_2$ equivalent emissions in both scenarios' (Mayers et al., 2015: 7) allowing us to subtract this figure from both the earlier ones to arrive at solely the distribution component they estimate. Downloading an 8GB PS3 game in 2010 therefore resulted in a range of **2.4 kg–8 kg** of $CO_2$ equivalent according to Mayers et al. (2015: 7).

Aslan's (2020: 196) data—consisting of a table and detailed descriptions of assumptions—provides a much clearer and far more up-to-date picture of where digital distribution emissions occur, as well as how much occurs at each stage:

- PSN store game retail 0.003 kg $CO_2$e/download
- CDN 0.051 kg $CO_2$e/download
- Access network 0.336 kg $CO_2$e/download
- CPE* 0.400 kg $CO_2$e/download (*Customer premise equipment— router, modem, etc.)
- Console download energy 0.034 kg $CO_2$e/download
- Console energy game file deletion 0.001 kg $CO_2$e/download (Aslan, 2020: 196)

Summing these components gives a figure of 0.825 kg for a digital download of a PS4 game, using a (2020 appropriate) average file size of 39.3 GB. Thus, even though game file sizes have more than tripled in the ten years since the earlier study, digital distribution emissions have reduced to at least one third, and potentially as much as a tenth of previous emissions. This can be explained as a product of increases in energy efficient internet infrastructure, greatly increased average download speeds which sees consoles spend less time 'on' while waiting for files to download, and lower overall emissions factors for power consumed. Aslan (2020: 196–7) emphasizes the role of energy efficiency improvements across console generations as well, with the conclusion that:

> power consumption has been reduced greatly between the PlayStation 3 case study in the focus of the Mayers et al. (2014) study and the PlayStation 4 model considered in this study (despite the current console having approximately ten times the performance).

This is an enormously encouraging finding, and if these trends continue points to real potential for emissions reductions. To this we can also add other secondary benefits, such as reduced plastic waste in landfill, and so on. To me this adds further weight that the future carbon neutral games industry will need to be primarily—if perhaps not *entirely*—a digital one. Aslan also discusses the other energy efficiencies in internet transmission in more detail elsewhere in his PhD, one of the other substantial factors contributing to a lower emissions profile from the shift in energy generation systems themselves. This energy system transition, already underway, is perhaps the single biggest factor in determining the future path of the carbon neutral games industry, and the goal of a 100% renewable energy system is now within sight.

In 2020, global emissions from power production accounted for as much as 44% of the world's total emissions, (CSIRO, 2021) and getting these to zero is a challenge that many leading scientists, engineers, and policy makers are already making great strides towards. As the proportion of global electricity generated through renewable technologies increases, the emissions associated with electrical energy and any given digital task—such as downloading a game over the PlayStation network—approaches zero.

Huge amounts of work, both theoretical and practical, have been done to demonstrate the feasibility of a completely renewable energy system in the past few years alone. An incredibly thorough literature review of over 60 different studies was conducted by Brown et al. (2018: 835) attempting to 'assess both the feasibility and the viability of renewables-based energy systems'. They describe many existing, mature technologies that some energy systems currently deploy, and which many in the near-future could, to ensure emissions free (or very nearly) electricity. They summarize their findings as follows:

> The technologies required for renewable scenarios are not just tried- and-tested, but also proven at a large scale. Wind, solar, hydro and biomass all have capacity in the hundreds of GWs worldwide. The necessary expansion of the grid and ancillary services can deploy existing technology. Heat pumps are used widely. Battery storage… is a proven technology already implemented in billions of devices worldwide (including a utility-scale 100 MW plant in South Australia and 700 MW of utility-scale batteries in the United States at the end of 2017). Compressed air energy storage, thermal storage, gas storage, hydrogen electrolysis, methanation and fuel cells are all decades-old technologies that are well understood. (Brown et al., 2018: 840)

In other words, 100% renewable power energy supplies are entirely feasible, and fast approaching a solved problem. The future for renewables seems to get brighter almost by the day, with many of the technologies mentioned above advancing and substantially reducing in cost even in the few short years since this review was conducted. From pumped hydro which can soak up and store excess solar during the day, storing it as kinetic energy to turn it back into clean hydro power once the sun has set, to more exotic time-shifting and power dispatching technologies like household, community, and grid-scale batteries, virtual power plants made from fleets of batteries which can export *en masse* to help stabilize the grid and make up shortfalls at peak times, and other solutions which have begun to enter the trial phase around the world. The authors also list a number of countries and regions with close to complete reliance on renewables already, demonstrating that not only are 100% renewable grids (or near enough) already possible, but they actually exist in some of the following places:

> Paraguay (99%), Norway (97%), Uruguay (95%), Costa Rica (93%), Brazil (76%) and Canada (62%). Regions within countries which are at or above 100% include Mecklenburg-Vorpommern in Germany, Schleswig-Holstein in Germany, South Island in New Zealand, Orkney in Scotland and Samsø along with many other parts of Denmark... There are also purely inverter-based systems on islands in the South Pacific (Tokelau and an island in American Samoa) which have solar plus battery systems. (Brown et al., 2018: 842)

Brown et al.'s (2018) paper however is just an analysis of technical viability—of whether or not 100% renewables are possible in practice. It is not possible yet to say that they necessarily *will*, though even the economics are pointing in this direction. There still remain substantial social and political obstacles in the way. In previous chapters we have already raised the ideological intransigence of climate skeptics, and hinted at the political-economic factors that prop up the prevailing fossil fuel regime, neither of which should be underestimated. There are also important social concerns with the way that renewable energy is financed and constructed, and who gets to reap the benefits, and whether those benefits are shared with the communities they are deployed within. For instance, the twinned 'duographs' of Howe's (2019) *Ecologics* and Boyer's (2019) *Energopolitics*, case studies of the intersections of wind power development with settler colonial logics, both document capitalist investment and local indigenous resistance in Mexico. Ketan Joshi's *Windfall* (2020) argues similarly for the importance of involving affected communities in the planning and the rewards of renewable infrastructure projects. Even with these issues, the underlying economics of power generation that simply require no fuels to generate are becoming irresistible even to capitalists themselves, with solar and wind now the cheapest forms of new power generation almost the entire world over.

The carbon neutral future of digital games distribution then will almost certainly need to align with digital downloading. For future researchers and others interested in conducting their own audit or analysis of digital distribution emissions, particularly how to reduce them, what are they key considerations and dynamics? From Aslan's study we learn that 'the size of game files and length of gameplay time were found to be key variables significantly impacting the results' (Aslan, 2020: iv). His analysis also aimed to put a figure on the emissions of various gameplay scenarios as well, using durations of 1 hr., 100 hr. and 1000 hrs, as well as different file sizes of 4, 10, 20 and 50 gigabytes downloaded, producing charts and scenarios for each. Changing these variables shifts the emissions balance and which methods his analysis identifies as having the lowest current footprint (disc, digital, or cloud-based gaming—see chapter 5 in Aslan (2020) for this full analysis). However, even this relies upon a fixed emissions profile, based on European electrical energy production, and as Mayers et al. (2015: 6) underscore, 'the validity of any analysis of this type deteriorates rapidly as time passes from the actual data of estimation or measurement.' In just a few short years then, as renewable continue to penetrate the market, these calculations may in fact shift further in favour of digital downloads to make them an inevitably greener option. Other parts of the world not covered by Aslan's (2020) analysis may be ahead or behind in the energy transition. But as we saw in the previous section of this chapter, sending discs to the other side of the planet carries a carbon cost that is stubbornly resistant to decarbonization, and may not be bearable for much longer. Replacement fuels for polluting transport industries are a long way off (save, perhaps for the electrification of trucking which is starting to accelerate—others are years, perhaps decades away).

Researchers and those looking to calculate the carbon footprint of specific digital distribution networks will do well to make sure their data is as up-to-date as possible, with accurate emissions factors for different electrical generation methods in the part of the world they are working on—a shifting target that also changes in real-time. Access to something like the UK's emissions intensity API which helpfully provides real-time breakdowns of emissions per kWh, even measuring regional differences, and other similar tools are increasingly available in other parts of the world to help real-time measurement of energy intensity. Likewise, many households now generate their own energy via rooftop solar, adding incentives to time-shift as much power use into daytime hours as possible (a trend that is becoming increasingly important in heavily solar endowed parts of the world, like Australia). During the day, for instance, on all but the cloudiest of days, virtually all of my household daytime energy needs are met by rooftop solar panels, with serious potential to reduce end user power consumption if power use can be avoided outside these periods.

There are also challenges in accurately measuring power usage inside data centers employed by digital distribution networks. Firstly, these facilities can remain technically beyond Scope 1 & 2 emissions boundaries if contracted through third parties. Even if platform holder decides to green their data center energy use, however there are real challenges to measurement of energy use and emissions. In one meta-review of data center energy usage, Dayarathna et al. (2015) surveyed over 200 different models of the estimated energy usage of different servers and their components, as well as cooling, networking and other powered systems that keep these highly energy intense locations operational. These locations are the energy intense heart of online activity, and deserve closer scrutiny. For example, Dayarathna et al. (2015) also include some staggering details about the scale of energy consumed by this part of the internet and its largely invisible infrastructure, noting that 'a typical data center may consume as much energy as 25,000 households. Data center spaces may consume up to 100 to 200 times as much electricity as standard office space... [and] the energy costs of powering a typical data center doubles every five years' (Dayarathna et al., 2015: 732). They offer a figure of 270 TWh of energy consumed by data centers annually in 2012, (Dayarathna et al., 2015: 732) and other estimates suggest total data center power consumption could be 1% of all the electricity used in the world even back in 2005. This figure comes from Andrae and Edler (2015: 133) who perform a macro-analysis of energy demands in computing technology, and predict that it could rise to 'around 3–13% of global electricity in 2030.' The contribution of these locations to overall energy demand underscores the importance of grid-scale transformation in power generation, which must be part-and-parcel with any digital transformation of the games industry, or else we risk undermining much of our efforts. Thankfully, in most parts of the world, power generation emissions are shifting in a positive direction, but this trend must accelerate and as discussed in the previous chapter, games companies can play a part in this.

What I have been able to offer here in this section is only a rough outline of what I believe is necessary for the decarbonization of the digital games industry. Others may find different points to emphasize in the analysis, or come to slightly different conclusions—however I strongly believe that it remains to be shown that a non-digital games industry can exist sustainably, and be carbon neutral. The burden of doing so rests with industry that wishes to continue it, and a shift in present expectations around modes of distribution is sorely needed. As Aslan (2020) has shown, detailed and careful work is required to produce a comprehensive and accurate picture of the carbon footprint of games distribution. Accounting for these emissions involves measuring aspects of different distribution platforms, data centers energy use, hardware efficiency, and accounting variations in local emissions intensity of power systems across the entire world, with much of this data currently privately held and undisclosed. If the emissions of game *development* are, as I claimed in the previous chapter, imminently addressable, the energy intensity of physical distribution remains much less so, complicated by geographical, economic and political challenges. For digital distribution, these challenges are not insurmountable, however. The original nature of digital distribution platforms as alternative approaches to DRM and issues of piracy have largely receded into the background today, having been overshadowed by the much more prominent role as a digital shopfronts, social networks, and providers of other gaming-related services. However these original impulses still influences the priorities of these digital platforms, and helps explain some of their persistent structural features and what may present challenges to efforts to decarbonize sales and distribution of games. In the most radical, idealized version of the future carbon neutral games industry, would we still accept the energy intensity associated with content delivery and other edge networks that push the physical location of servers closer to end users? If we prioritized lower emissions over the speed and convenience of high-speed networks, would the digital distribution architecture still look as it does today? And are these sorts of tradeoffs even necessary if we can achieve a 100% renewable grid? Many questions remain.

## CONCLUSION

In this chapter we have looked at what it takes to get games into the hands of players, whether on a disc or by digital download. We have looked at the fairly slim yet extremely important literature on the emissions of disc-based distribution, and my own back-of-the-envelope estimates extending calculations from known European emissions to more far-flung regions of the world, like my home country of Australia. The result of this desk analysis—which still requires real world validation through a full life-cycle assessment—gives a sense of the approximate emissions from disc-based distribution, and some sense of their potential scale. Despite their relative modesty in the overall global emissions context, in an absolute sense these emissions cannot be ignored if we are committed to the carbon neutral games industry—as we must be. These are emissions the games industry and *no one else* is responsible for. If it does not do something about them, then no one else will. My principal suggestion would be to focus on accelerating digitization. In the short term, offsets can be used. Longer term, solutions to plastic waste and the embodied emissions in digital discs and cartridges demand a more lasting solution.

Indeed, if we considered only the existing analyses on current emissions, digital distribution might seem to be no solution at all. If things remained as they were in Mayers et al.'s (2015: 12) initial analysis of digital distribution in 2010, 'Thankfully, much has changed since, and energy efficiency is now front and center. Aslan's (2020) updated modelling of internet infrastructural power consumption found that even with the tripling of game file sizes in the decade or so since, there are many scenarios in which digital download emissions are less than disc-based emissions. For large downloads Aslan (2020) found some cases where these emissions still surpassed disc distribution (at least in Europe) however we should not become fixated on a single, static analysis and instead look beyond and to the future, taking note of the already clearly trending-lower emissions intensity of electricity generation. Many games are almost already constantly updating, with new downloads for content and patches on a routine basis. The infrastructure and systems are already in place to enable greater digitization. According to Aslan (2020) gaming's power use at runtime is approximately 60% of the overall emissions footprint on a per-gameplay-hour basis. Gameplay thus dwarfs the footprint of whatever game acquisition method one uses (Aslan 2020: 194–198) though it does not render distribution considerations irrelevant. As a result of this proposal, however, hard times may be coming for the physical games retailer, despite the heroic rallying of GameStop's fortunes we saw following the attention of r/WallStreetBets in early 2021. Planning for how to ensure a just transition for workers in this sector should also begin now, just as much as the platforms themselves should experience a reduction in their monopoly powers.

The games industry will need to consider what is necessary and adequate for the goal of full decarbonization. My analysis of the embedded carbon costs involved in the distribution of discs to Australia adds further weight to the position that disc-based games distribution is incompatible with this goal. Zero carbon electricity generation is already plausibly within reach, in a growing number of parts of the world, and must become a central pillar in any decarbonization strategy of the digital games industry. Unlike the global shipping and trade system, for which complete decarbonization remains a distant prospect, electricity generation systems from domestic rooftop solar to large scale wind and solar are already beating many existing fossil fuel generators on price of power. A fully digital games industry is one that is much more readily positioned to benefit from the transition to renewable energy generation as it gathers momentum.

The goal of this chapter has been to prompt readers—from developers, to players and industry decision-makers—to consider this usually hidden, and often ignored part of the digital games industry, and start a conversation about ways of addressing the issue of emissions in distribution. For the sake of our planet, and for the sake of ourselves, I hope this can begin a discussion between publishers, developers and players about what a truly sustainable, truly ecological games distribution system might look like and how we get there.

## References

Ahmad, D. (2020, July 31). EA said that 52% of its console full game unit sales in the past 12 months were via digital download. *Twitter*. https://twitter.com/zhugeex/status/1288948030604009482. Accessed 1 Apr 2021.

AirBridgeCargo. (2020). AirBridgeCargo Airlines – Boeing 747-400F. https://www.airbridgecargo.com/en/page/38/boeing-747-400f.          Accessed 1 Apr 2021.

Andrae, A. S. G., & Edler, T. (2015). On global electricity usage of communication technology: Trends to 2030. *Challenges, 6*(1), 117–157.

Aslan, J. (2020). *Climate change implications of gaming products and services.* PhD dissertation, University of Surrey.

Barlog, C. (2018, March 22). God of War has gone gold. A message from Corey Barlog. *PlayStation Blog.* https://blog.playstation.com/2018/03/22/god-of-war-has-gone-gold-a-message-from-cory-barlog/. Accessed 1 Apr 2021.

Boyer, D. (2019). *Energopolitics.* Duke University Press.

Brown, T. W., Bischof-Niemz, T., Blok, K., Breyer, C., Lund, H., & Mathiesen, B. V. (2018). Response to 'Burden of proof: A comprehensive review of the feasibility of 100% renewable-electricity systems'. *Renewable and Sustainable Energy Reviews, 92,* 834–847.

Centre for Aviation. (2021). Vienna International Airport. *centreforaviation.com.* https://centreforaviation.com/data/profiles/airports/vienna-international-airport-vie. Accessed 1 Apr 2021.

CSIRO. (2021). *Global carbon dioxide emissions.* CSIRO. https://www.csiro.au/en/research/environmental-impacts/emissions/global-greenhouse-gas-budgets/global-carbon-budget. Accessed 1 Sept 2021.

CyberpunkGame. (2020, October 5). Cyberpunk 2077 has gone gold! 🎮 See you in Night City on November 19th! *Twitter.* https://twitter.com/CyberpunkGame/status/1313067011455569921. Accessed 1 Apr 2021.

Dayarathna, M., Wen, Y., & Fan, R. (2015). Data center energy consumption modeling: A survey. *IEEE Communications Surveys & Tutorials, 18*(1), 732–794.

Donovan, T. (2010). *Replay: The history of video games.* Yellow Ant.

Edwards, C. (2013, November 5). Valve lines up console partners in challenge to Microsoft, Sony. *Bloomberg.* https://www.bloomberg.com/news/articles/2013-11-04/valve-lines-up-console-partners-in-challenge-to-microsoft-sony. Accessed 1 Apr 2021.

Gillespie, T. (2010). The politics of 'platforms'. *New Media & Society, 12*(3), 347–364.

Howe, C. (2019). *Ecologics.* Duke University Press.

Howitt, O. J. A., Carruthers, M. A., Smith, I. J., & Rodger, C. J. (2011). Carbon dioxide emissions from international air freight. *Atmospheric Environment, 45*(39), 7036–7045. https://doi.org/10.1016/j.atmosenv.2011.09.051

IPCC. (2014). In O. Edenhofer, R. Pichs-Madruga, Y. Sokona, E. Farahani, S. Kadner, K. Seyboth, A. Adler, I. Baum, S. Brunner, P. Eickemeier, B. Kriemann, J. Savolainen, S. Schlömer, C. von Stechow, T. Zwickel, & J. C. Minx (Eds.), *Climate change 2014: Mitigation of climate change. Contribution of working group III to the fifth assessment report of the intergovernmental panel on climate change.* Cambridge University Press.

Johns, J. (2006). Video games production networks: Value capture, power relations and embeddedness. *Journal of Economic Geography, 6*(2), 151–180.

Joseph, D. (2018). The discourse of digital dispossession: Paid modifications and community crisis on steam. *Games and Culture, 13*(7), 690–707.

Kirkpatrick, G. (2012). Constitutive tensions of gaming's field: UK gaming magazines and the formation of gaming culture 1981–1995. *Game Studies: The International Journal of Computer Game Studies, 12*(1), 3.

Mayers, K., Koomey, J., Hall, R., Bauer, M., France, C., & Webb, A. (2015). The carbon footprint of games distribution. *Journal of Industrial Ecology, 1*, 402–415.

Moore, C. L. (2009). Digital games distribution: The presence of the past and the future of obsolescence. *M/C Journal, 12*(3). https://doi.org/10.5204/mcj.166

PWC. (2020). *Outlook: The Australian Entertainment & Media Outlook 2020–2024.* https://www.pwc.com.au/industry/entertainment-and-media-trends-analysis/outlook.html. Accessed 1 Apr 2021.

Sea-Distances.org. (2021). *Sea Distances shipping calculator.* https://sea-distances.org/. Accessed 1 Apr 2021.

Sony. (2020). *Sustainability Report 2020.* https://www.sony.net/SonyInfo/csr/library/reports/SustainabilityReport2020_E.pdf. Accessed 30 Mar 2021.

Srnicek, N. (2017). *Platform capitalism.* Wiley.

Taffel, S. (2012). Escaping attention: Digital media hardware, materiality and ecological cost. *Culture Machine, 13*.

Toivonen, S., & Sotamaa, O. (2010). Digital distribution of games: The players' perspective. In *Proceedings of the international academic conference on the future of game design and technology* (pp. 199–206). ACM.

van der Loeff, W., Schim, J. G., & Prakash, V. (2018). A spatially explicit data-driven approach to calculating commodity-specific shipping emissions per vessel. *Journal of Cleaner Production, 205*, 895–908.

VGChartz. (2021). Platform totals. *VGChartz.com.* https://www.vgchartz.com/analysis/platform_totals/

World Shipping Council, International Maritime Organization. (2020). Industry issues: Carbon emissions. *Worldshipping.org.* https://www.worldshipping.org/industry-issues/environment/air-emissions/carbon-emissions. Accessed 1 Apr 2021.

GAMES

FromSoftware, Inc. (2016). *Dark Souls 3.* FromSoftware and Bandai Namco Entertainment.

Square Enix Business Division 1. (2020). *Final Fantasy VII Remake.* Square Enix.

# CHAPTER 6

# The Carbon Footprint of Playing Games

This chapter is all about playing games and the amount of energy that is entailed, an essential part of the decarbonization picture given the apparently unceasing growth in game players around the world. This chapter first provides reference figures for different game consoles energy use, combining real-world measurements of typical power draw, where possible, and where it is not I provide reference to the rated power values of the console's power supply. The goal here is primarily to provide a simple, rough figure for developers and anyone else who wishes to estimate the energy used by players. The chapter discusses a methodology for estimating player emissions used by Space Ape Games in their industry leading offsetting efforts, with developers uniquely placed to collect usage metrics that allow much more accurate estimates than possible previously. Previously, we have had to rely on time-of-use studies of various consoles and player usage habits to estimate footprints. I add my own real-world measurements of the launch version of the non-optical drive PlayStation 5 based on my own console's performance during real-world use, adding detail to the contemporary emissions picture. With compelling estimates for the scope of emissions associated with gaming devices power consumption, the second section of the chapter focuses on energy efficiency, and the case for what can be gained from improved efficacy of gaming devices in the form of avoided emissions.

© The Author(s), under exclusive license to Springer Nature        149
Switzerland AG 2022
B. J. Abraham, *Digital Games After Climate Change*, Palgrave
Studies in Media and Environmental Communication,
https://doi.org/10.1007/978-3-030-91705-0_6

In the final section of the chapter, I consider what game developers might do to reduce these energy demands, whether by changing the kinds of games that are made, with different kinds of games leading to different kinds of energy use. Much of these ideas however face substantial unpopular trade-offs, or are predicated on a kind of climate ascetism which is unlikely to appeal to the wider game playing audience. I argue, however, that even if we do not face the kind of resource scarcity presumed by the need to reduce the energy demands of gaming, there remains some generative potential in thinking about and designing for these sorts of constraints. The games industry has a history, however, of flourishing within constraints, especially hardware constraints. Less important perhaps than whether an energy-poor future eventuates, I suggest that a focus on doing more with fewer hardware upgrades will almost certainly become a part of the games industry's future given the resource intensity of this aspect of games.

The energy use and emissions

intensity of player consumption is the final part of the direct emissions footprint for the games industry. Of the three categories of emissions discussed thus far (development emissions, distribution emissions, and play emissions) this last one represents the biggest portion of the emissions picture of the current games industry, but also a hard challenge for the industry itself to address. Firstly, there is a great challenge in assembling an accurate picture here that lies in the huge variability in patterns of use of different devices, with different play routines, and different energy usage profiles of each hardware device leading to substantial differences in emissions, across an enormous diversity of existing gaming devices. Secondly, there is the same complexity of regional energy generation and emissions intensity that we have seen time and again in previous chapters, with the emissions of power generation by no means uniform across the world. The emissions of play is, however, one area where platform owners like Microsoft, Nintendo and PlayStation have more leverage than software developers. Hardware designers can use their control over this process to further prioritize energy efficiency in hardware revisions, as we see through Aslan's (2020) analysis. Platform holders are also in position to publish anonymous usage metrics helping greatly to clarify the picture around emissions. There are some encouraging signs indicating that some might already be considering this, but until then individual developers can greatly help by collecting data themselves—with Space Ape Games demonstrating how this might be done and how it is relatively feasible (for some) to offset player's emissions once measured accurately.

Complete decarbonization of play however is most likely going to be much more complex and difficult to achieve than those reductions described in the preceding chapters. The changes necessary to games themselves to substantially reduce energy use remaining unclear as to just how much of a return on effort they stand to produce, or even how widely they could be accepted. To this we can add the same observation about emissions from electricity reducing drastically with the transition to renewables, which might lead us to ask why we would even bother, at least in the long term. Therefore once again the emissions of play are greatly determined by and inseparably tied to the larger political and economic struggle to decarbonize global energy systems.

## Console Energy Use & Carbon Emissions

By this point, readers may have some sense of the complexities involved in arriving at accurate emissions figures in a range of areas, and this applies to playing games as well. Game consoles, iPhones, PCs and other digital devices used for gaming don't always use a constant amount of power, with watts drawn fluctuating in response to the activation of different components, different levels of hardware utilization, the availability of power saving modes, and so on. The maximum rated power draw of a device's power supply can give some indication of how much it uses at peak performance, but hardware rarely runs at perfect maximum, experiencing minor fluctuations and deviations in response to what is required of it, dropping when the pause menu is pressed, or when the graphical intensity of a scene changes. Beyond power usage, the emissions from a unit of power consumed can vary, sometimes quite dramatically with differences in local power generation emissions fluctuating in real time, across the day, and as part of longer, seasonal trends. The same device in one household might have a completely different pattern of energy consumption in another, based on the different whims and habits of its owners. One household might have solar panels and a battery enabling it to draw very little from the power grid. Others might only have grid power. These factors mean that, in the absence of direct measures of usage and the availability of real time emissions factors, once again broad trends and rough estimates are the best we can probably hope for.

A calculation of power consumption involves the multiplication of two measures: power level (in watts) over a given time period (usually hours). For example, if a fridge uses a constant 1000 watts over one hour, we say that it has consumed 1 kilowatt hour (kWh). This is an easy calculation when power used remains constant, but as mentioned many devices draw variable amounts of power. For an accurate measure we can use a device known as a power meter to add up consumption as power levels rise or fall over time. For an estimate of a large number of game consoles we can use an average power figure, combined with the duration players typically use their gaming device. Delforge and Horowitz (2014) conducted just such a study, looking at the Xbox 360 and the duration the console spent in various modes for a large number of users. Though it was completed several years ago, with consoles two generations old, it at least provides an indication of the sorts of power drawn by the device, and more importantly the patterns of time it was used for, with figures for standby, media streaming, and gaming modes. In the absence of the ability to capture metrics in real time, we can use a method that takes 'average' power levels and keeping things a bit simpler, accepting some inaccuracy as a trade-off. When completed, all we then need is to take the kWh figure and multiply it with the emissions factor for a given location to allow developers and other interested parties to estimate players emissions. It should be noted that Space Ape Games, as first raised in Chap. 4, have been the pioneers at this, and their methodology remains the closest thing to industry best practice. Their report on their player's 2019's emissions went to the trouble of calculating and then offsetting all of them for the year, one of only a few gaming companies that have done so (see Chap. 4). With current and up-to-date emissions factor figures, and an expanded list of devices, this same method can be applied to game by its developer to estimate player's carbon emissions.

The list of devices and their power usage figures is as follows:

### *PlayStation*

PlayStation 3 (2006/7)—400 W-200 W (rated PSU)
PlayStation 4 (2013/4)—137 W-78 W (measured by Aslan (2020: 232))
PlayStation 4 Pro (2016)—280 W (rated PSU)
PlayStation 5 (2020)—250 W (measured peak; 190–220 average; non-optical drive model), 350 W (rated PSU)

### *Xbox*

Xbox 360 (2005)—200 W–120 W (rated PSU)
Xbox One (2013)—220 W (rated PSU)

Xbox One X (2017)—245 W (rated PSU), 177 W (measured by Eurogamer/Digital Foundry (Leadbetter 2020))
Xbox Series S (2020)—120 W (rated PSU)
Xbox Series X (2020)—315 W (rated PSU), 211 W (measured by Eurogamer/Digital Foundry (Leadbetter 2020))

*Nintendo*

Nintendo Wii (2006)—45 W (rated PSU)
Nintendo Wii U (2012)—7.5 W (rated PSU)
Nintendo Switch (2017)—39 W (rated PSU)

Some further explanation of these figures is required. Where a range is included, this is to account for the power usage across different hardware revisions of a console—as typically newer console revisions became more efficient over time. The "rated" figure is either one supplied by online parts sellers or derived from the voltage & current specifications of the Power Supply Unit (PSU) (Watts = Volts x Amps), and which typically represents greater capacity to supply power than the device needs (hence, the 100 watt disparity between the maximum power usage I measured with my own PS5 and what the PSU is theoretically rated for). Space Ape Games outline a method for taking power consumption figures like these and combining them with duration-of-use figures (they used internal metrics built into their games). They offer the following explanation of how to achieve an end user emissions figure, as simple as multiplying the right numbers together:

> For player usage, we combined our own internal analytics data which details the number of hours played per player in each country with a table of emissions data for electricity across the world, published by Ecometrica. (Space Ape Games, 2020)

Digging into the figures for one country shows what is involved. Space Ape reported in their spreadsheet (linked from the online report) that the UK player-base played around 4.4 million hours of their games collectively in 2018. This was done (we can safely assume) across a variety of different mobile devices. To achieve an estimate of these emissions attributable to play, the total hours played figure is multiplied by, first, the estimated power (in Watts) assumed to be used by a phone playing their game. This is multiplied by the duration of play, becoming a number in Watt-hours, which is then divided by 1000 to become kilowatt hours (kWh). Space Ape Game's publicly available spreadsheet of this work provides an estimate of 9265 kWh of power used by players in the UK across the whole of 2018. This kWh figure is then multiplied by an emissions factor of 0.548402315 (representing a measure of how may kg's of $CO^2$ equivalent is emitted per kWh of power generated—itself an average figure, as emissions factor changes in real-time) arriving at a figure of 5081 kg of $CO^2$ equivalent emissions. In simpler terms, Space Ape Games players in the UK emitted around **5 tonnes** worth of $CO^2$ through the power used by their phones playing their games. These 5 tonnes are just one country worth of emissions, and when the same exercise is repeated across the rest of the world, the total was just over **181 tonnes**. At the end of 2018, the EU $CO^2$ price was around €20–25 per tonne allowing the company to purchase a modest carbon offset worth about €3000–4000 in order to offset the entirety of their players emissions.

There are a couple of complications, however, that affect this estimate, and as always it is important to remember that these are estimates. Firstly, and in response to this very fact, Space Ape Games themselves recommend building padding into the estimates, achieved partly by using emissions figures from 2011, and which were higher than probably necessary. As we saw, the 2018 calculation for UK players used a $CO^2$ equivalent emissions factor of 0.548402315—a bit over half a kilogram of $CO^2$ equivalent emissions for every kilowatt hour generated. However, the UK Govt's own figure for its national emissions factor for 2018 is not quite half the one used by Space Ape: only 0.28307. Space Ape Games notes in their guide that they have built in a substantial degree of 'pessimism' into all their projections, attempting to over-estimate emissions rather than under count. This is worthwhile, with few downsides to offsetting more than required (except cost).

The second complication however is in the assumption of a flat energy use figure for mobile phones (in Watts), which the report notes is based on research done by Malik (2013) who provides a figure of 2.0885 Watts used by most smartphones. The differences in power usage between different mobile phones are much less than even between consoles for instance, but even so, given the huge number of hours played, even small changes in the power consumed has an outsized effect on overall emissions. Other researchers have found figures that are quite a bit higher than Malik's (2013), with Kim et al. (2014) reporting power consumption of an unspecified game reaching almost 4 Watts. The lack of clear figures for smart-phones (not to mention other gaming devices) underscores the importance of greater transparency around the real-world power consumption of devices—an area platform holders and hardware manufacturers can be active. Given current trends toward more powerful mobile phones with bigger capacity batteries these figures could also trend upwards over time. Power benchmarking on Apple's A13 chip (powering the iPhone 11) peaked at around 5 Watts in some tests (Frumusanu, 2019) more than double the figure provided by Malik (2013). It's still possible that Space Ape Games original figure of around 2 Watts is the better reflection of real-world use, depending on how much CPU/GPU utilization a particular game requires, and again underscoring the impossibility of perfect accuracy. Collecting information on user's handset models and modelling a particular game's power draw in real-world conditions would add further accuracy, though at perhaps the cost of introducing additional data governance and privacy issues. In any case, for now (and only as long as energy systems remain non-renewable) estimates, a sort of 'close enough' figure with built in pessimism and extra padding, remain a safe choice, still far better than nothing. Over time, the accuracy of this kind of modelling should improve, and is certainly highly desirable for both cost effectiveness of offsetting efforts and confidence that these figures represent actual emissions.

One last factor, however, that does not appear to be accounted for anywhere in Space Ape Games calculations, however, is derived from the peculiarities of smartphones and devices with internal batteries. There is a gap between the *time* of generation of emissions and when that power is used by the device as it draws on its internal battery, as well as another gap between the power used by the device and the power drawn from the grid to charge that battery. Despite the overall quantities being relatively small, both the charging process itself, and the conversion of chemical potential stored in the battery back into electrical energy for the device are not perfect, and losses occur with imperfect conversion from AC to DC especially during charging (Kim et al., 2014).

Here we can draw on other manufacturer's specifications, like Apple's sustainability report for devices which includes data on charger efficiency. For comparisons sake, the charger that came with my iPhone 7 Plus (which I am still using many years later) suggests that 73–74% of the power that is drawn from the grid ends up in the battery itself (Apple, 2017: 2). This indicates that a substantial portion is lost, and that the actual power drawn from the grid (which is where the emissions are created) is higher than the measured power use from the phone itself. We can account for this by increasing the energy use figure by this proportion—giving a final emissions figure that is just shy of 30% higher than previous. Given the small amounts of power involved here, these charger losses only matter when scaled up to millions of players, and smart phones are already some of the most energy efficient gaming devices around. The final cost of emissions does not increase substantially by incorporating these refinements—a best-guess might be perhaps a new figure of **245 T of $CO^2$-equivalent** emissions vs the previous figure of 181 T. Given the EU carbon price above, this adds only about another €1600, and given Space Ape already built in 'padding' with a generous overestimate in the emissions factor figure, even this is probably unnecessary in this case.

A similar generosity in calculating and offsetting player emissions is also recommended by the only other game developer I am aware of that has begun to do so. Rovio's (2020: 7) 2019 CSR report notes the following to explain why they recommend similar over-estimation:

> Recognizing the fact that Rovio players are interacting with their mobile devices in other ways than just playing Rovio games, Rovio decided to compensate for a full charge to keep the methodology simple and ensure Rovio would not under-compensate the impact. In total, this equated to 19,476 carbon tons ($CO^2e$ / year). The full amount was offset via a United Nations Carbon offset platform.

Though seeming to use a different calculation methodology, what I suspect they are suggesting here is that whenever player-hours would be less than a 'full charge' of a device, they have rounded up. Both the simplified Rovio method and the Space Ape method are broadly applicable to other games and game developers, provided the equations are updated to reflect the equivalent device, that accurate play-duration figures are collected, and up-to-date emissions factors are available.

If we were to change the device being used by players to a console, however, the picture changes dramatically. Replicating and slightly extending Aslan's work to the latest console generation I undertook a short period of measuring the power consumption of my own PS5 shortly after obtaining a console at launch. To measure my PS5's power draw, I used a simple socket-based power meter to collect the total power consumption over two weeks in December 2020. I left the console plugged in to the power meter and went about playing games as usual. Every so often I looked at the peak power that had been drawn, and confirmed a fairly consistent power draw (as per the figures listed earlier). Despite the much higher performance of the PS5, thankfully its power consumption has not increased by the same proportion. This reflects Aslan's (2020: vi) findings (referring to the difference from PS3 to PS4) that 'show that a decoupling of performance and power consumption has been achieved for the first time between successive consoles platforms', which seems to have also occurred in the PS5. During the two weeks, I played mostly games like *Assassins Creed: Valhalla* (PS5 mode) and *Destiny 2* (in PS4 compatibility mode). The result was that my PS5 used about 8kWh of power over that duration, or around 0.5 kWh per day, which seems to reflect the peak power usage of around 250 watts reported by various outlets and my own sense of how long I played for, and the more average reading closer to around 190–220 watts (I estimate that I played around 3 hours a day). The model I tested it on, was the PS5 SKU without an optical disc drive, though this likely made a very small difference if any. A daily average of 0.5kWh would represents fairly heavy usage, and may seem a lot but I play quite a lot of *Destiny 2*, logging on most evenings for a few hours. In general, my own measurements lined up with the time-of-use figures expected from the higher-end usage case of Aslan's own study.

With gaming consoles using up to 100 times more energy than a smart-phone, this has a similar sized impact on emissions calculations. According to Aslan (2020: 232), different iterations of the PS4 over its lifetime used as little as 73 W or as much as 130 W, and my own measurements take the PS5 even higher. If Space Ape were offsetting the emissions of its end users for a PS4 game instead of a mobile game, and had the same hours played in the UK as their mobile games in 2018, they would be attempt-ing to offset around 60 times the emissions. That takes the offset amount for just the UK to over **300 T CO²e**, and the world's total to **10,860 T CO²e**. The offsetting cost of this much $CO_2$, even at the lower end of the EU carbon price (just €25 per tonne), becomes an intimidating €271,500. With carbon prices increasing further still, reaching €50 per tonne by early 2021, and projected to go on increasing to as much as €90 by 2030, it seems unlikely that all but the most committed—or the most profitable—game development companies will be able to find room in the budget for offsetting console-based gaming emissions on this scale.

Similarly, Aslan's (2020) research put a figure on the total energy used by the entire PS4 European install base over its lifetime. To arrive at total power consumption figures, Aslan (2020: 63) canvasses a range of time of usage statistics, drawn from market researchers like Nielsen and others, arriving at 'an estimate of 1.79 hours per day' with a final estimate for 'total on time for PlayStation 4 [at] 2.24 hours per day' (Aslan, 2020: 63). To increase the accuracy of the estimate, he examined a range of different hardware revisions of the PS4 and their differing power consumption, from the earliest (and least efficient) to the most recent 2017 revision which reduced overall power consumption by as much as 60 watts. Aslan (2020) then added a detailed analysis of the sales figures and lifetime of each particular SKU, and estimated power usage on a monthly basis, using direct measurements of power usage for both gaming and other modes of operation. He concluded that:

> the actual PlayStation 4 power consumption today is lower, for all modes; than previous researchers had predicted, in fact power consumption has been reduced to levels lower than had been predicted for 2019. (Aslan, 2020: 85)

Despite this, and partly due to the huge success of the PS4 console, 'life-time cumulative electricity use of consoles in Europe…is estimated at 13 TWh…. if high estimates for console usage (4.4 hours per day) are representative of actual usage, then lifetime electricity use could be as high as 27 TWh in Europe' (Aslan, 2020: vii) (One TWh is one billion kWh). When combined with average world emissions factor of 0.689697213 (again a crude estimate), this gives a total figure of **18.6 million tonnes of $CO^2e$** emissions for the entire European PS4 install base—or €372 million worth of carbon emissions (even at just €20 a tonne). Could the games industry even survive if it had to pay the true cost of these emissions?

So far, this has only concerned console and mobile gaming, but an important contribution comes from the work done by Mills et al. (2019: 172) in their groundbreaking study of gaming devices energy use, particularly focussing on gaming PCs (circa 2016) which are typically harder to estimate energy use for given the staggering variety of hardware configurations possible. From these tests, which focused on low, mid, and high-spec PCs, also including games consoles, VR, and streaming game services, they arrive at the following estimate for the total US gaming market:

> The 134 million gaming platforms existing in the U.S. as of 2016 consumed 34 TWh/year, corresponding to a $5 billion/year expenditure by consumers, and 24 million tons $CO^2$-equivalent/year of greenhouse-gas emissions (equal to the emissions of 5 million typical passenger cars). (Mills et al., 2019: 172)

This US figure of 24 million tonnes per year is roughly similar to my esti-mate of the scale of the industry's emissions from game development dis-cussed in Chap. 4. Despite the higher per-device energy use of gaming PCs, Mills et al. (2019: 172) noted that 'consoles were responsible for 66% of the total energy use in 2016 for computer gaming across the duty cycle, followed by 31% for desktops, 3% for laptops and less than 1% for media-streaming devices.' Clearly the scale of gaming happening on con-soles accounts for much of this footprint even if their per-device intensity remains lower, reinforcing the importance of a focus on the efficiency of consoles.

Given all the challenges outlined in this section regarding power usage, player usage, emissions factors, energy efficiency, and so on—Space Ape Game's accounting project remains an invaluable tool and case study for game developers. It remains the most comprehensive attempt at measur-ing and offsetting players' emissions to date, and offers a compelling road-map for measuring and offsetting emissions in the games industry—even if not all may be able to replicate their full achievement. The methodology and documentation of it shared online remains a great example for other developers to follow, and there should be no reason that other game development studios could not build in the same usage statistics and fol-low suit. Even if only to produce estimates of their player's current footprint.

Given the cost of offsetting these emissions, once they are no longer confined to playing with devices that are amongst the world's most energy efficient, then it seems that offsetting player emissions for Triple–A game play may not be feasible at present. According to standard GHG account-ing protocol these emissions are ultimately the responsibility of players themselves, and there may be a case to be made for great consumer aware-ness and action. The figure of multiple hundreds of millions of euros to achieve the offsetting necessary for a carbon neutral PS4 represents a cost that would radically impact the sector's bottom line. It represents a sub-stantial imposed cost in the already eye-wateringly expensive Triple–A game sector, that would likely be passed on to consumers in the form of higher prices in any case. A serious reconfiguration of the existing eco-nomics of game development might even be required if companies wish to take this on more broadly, with almost no telling how this might shift game playing patterns, the class makeup of players, and so on. A partial solution in the interim, then, until the realisation of fully fossil fuel free power worldwide may be a greater focus on efficiency.

## THE IMPORTANCE OF ENERGY EFFICIENCY

Though it might seem far-fetched now, it's certainly plausible that the coming decade or two sees a radically different attitude towards what is permissible from the induced energy demand of hardware and software platforms. Regulators looking to make quick essential emissions cuts could choose to target luxuries and non-essential computation, like games, cryptocurrencies and the like, or mandate targets in energy efficiency increases. Knowing the source and scale of emissions intensity entailed by the whole of the games industry can help decisionmakers head off some of these challenges, and allow for preemptive planning before any of these sorts of regulations or taxes are imposed. Consider, for instance, the European Union's "Carbon Border Adjustment Mechanism" agreed upon in July 2021 which essentially applies carbon tariffs to carbon intensive imports of cement, iron and steel, aluminium, fertilizers and electricity (European Commission, 2021). This same mechanism could quite feasibly be expanded to include other items, from consumer goods that induce significant power demand beyond agreed maximums, or that represent substantial embedded carbon emissions in the device itself like high tech and gaming devices. On July 1, 2021, the second tranche of a suite of Californian regulations came into force that affect high power consumption computing devices. The legislation demands a certain 'expandability score' (i.e. future-proofing) of devices, and for those that do not meet certain scores, are limited to consuming a certain amount of power per annum, or are banned from sale (Claburn, 2021). Surely this is a sign of things to come, as publics and governments both demand greater action.

Alternatives will need to be imagined, developed, and implemented, and the energy efficiency initiatives for the newest generation of consoles made a priority by both Sony and Microsoft. This is achievable, as Aslan (2020: vii) puts a figure on the energy saved by efficiencies between PS4 hardware revisions at 'between 6 to 8 TWh of avoided electricity consumption.' Other initiatives like rest mode and other low-power functions add up to even greater avoided emissions:

> the avoided electricity use, as a result of the energy efficient technologies adopted in PlayStation 4, is estimated to be 30 TWh, equivalent to the annual electricity production of Denmark. (Aslan, 2020: 91)

These figures show the impressive scale of the energy involved with keeping the games industry going in its present state, as well as the potential to save a huge amount in emissions through even relatively small increases in energy efficiency. The World Energy Council noted in its 2016 report on the transition away from fossil fuels that 'energy efficiency improvements over the last 15 years saved the world 3.1 Gt of energy and 7 Gt of $CO^2$, which corresponds to 23% of global energy consumption and 21% of global $CO^2$ emissions in 2015' (WEC, 2016: 4). These efforts should clearly be supported, deepened and accelerated as much as possible in games as well.

Some readers might raise objections to a focus on gaming hardware energy efficiency, given the so-called 'Jevons paradox'—the supposition that increases in energy efficiency are cancelled out by a 'rebound effect' either in the wider economy or in sector-wide productivity that ends up increasing overall energy consumption. However, in his review of the literature on this phenomenon and on rebound effects in the economy more widely, Sorrell (2009: 1467) concludes that the case for the so-called paradox 'relies largely upon theoretical arguments, backed up by empirical evidence that is both suggestive and indirect.' In other words, there is a lack of unambiguous and empirical evidence for the existence of this paradox. It is also clear from Sorrell's (2009: 1467–8) analysis that it is especially unsuited to explaining or predicting the outcome of energy efficiency initiatives in electronic devices, concluding that the:

> Jevons' Paradox seems more likely to hold for energy-efficiency improvements associated with the early stage of diffusion of 'general-purpose technologies', such as electric motors in the early twentieth century. It may be less likely to hold for the later stages of diffusion of these technologies, or for 'dedicated' energy-efficiency technologies such as improved thermal insulation.

Improved energy efficiency of gaming devices seems an extremely unsuitable candidate to point to possible 'rebound' effects negating efficiency increases—the price and availability of power having little to do with how much gamers play games, which is more a function of lifestyles and habits than anything else. Similarly, games consoles are unlikely to present a significant enough part of global energy consumption sufficient to shift the availability and prices of electricity—at least, not on their own. I have taken pains to repeatedly emphasize throughout this book, however, that efforts must also be directed at supporting the transformation of grid scale electricity generation to completely emissions free electricity. The completion of this transition, however, ends up diminishing the importance of efficiency improvements as well—the emissions of game play at that point becoming nearly zero. This new hypothetical paradox, however, should not deter us as this situation is likely a decade or more away. Alone, individual game development studios may be able to implement some power-saving features into their games—and in the next section I discuss some opportunities for these and other kinds of changes—but the greater impact is to be made as an industry, both through changing the design of devices, and through its substantial lobbying power and corporate influence. To me, the games industry seems to stand a better chance of having impact if it unambiguously takes the sorts of actions suggested in Chaps. 4 and 5 to promote this transition, with energy efficiency efforts a slightly lower priority. Useful, especially in the next few years until renewable power is all but ubiquitous, but not a silver bullet.

Similarly, while improving the energy efficiency of devices is important, it is not the sort of action that is within reach of most game developers. However there are other ways to have an impact on the emissions of end users by changing the *kinds* of games that get made. As mentioned previously, the power draw of particular devices can vary quite dramatically depending on a variety of factors like hardware utilization, graphical intensity and so on. There are also other emissions—like the emissions embedded in new or upgraded devices—which can also be targeted through changing design patterns. If we agree that the goal *must* be a climate neutral games industry, then reducing Scope 1 and 2 emissions—that is the emissions from energy used by game developers themselves, either with offsets or fully renewable power, must remain the top priority and the first steps that any game developer or game development company takes towards this goal. These emissions are the most direct immediate responsibility of game developers, and conveniently some of the simplest to measure and the quickest to ameliorate, with additional benefits from reducing consumption such as saving money, and so on. But after prioritizing this more substantial work, there is also a degree to which game developers can participate in the reduction of demand for energy downstream, through the games that they make. This too could become a site for action in the games industry that is attendant to the challenges of climate change, one that wants to produce ecological games.

## GAMES AFTER CLIMATE CHANGE

According to the GHG emissions protocol accounting standard, downstream player emissions are technically an optional responsibility of the game maker. However there is nothing stopping a studio, solo developer, or even a publisher taking an interest in and even actively designing for energy efficiency through games themselves. It is also possible, in some of the more extreme climate scenarios, that our future holds some level of energy scarcity or other constraints that block consumers access to power, whether through emissions curtailment, or perhaps via draconian climate interventions, the simple need to respond to the intermittent nature of renewable generation or fluctuating prices, or by the impacts of extreme weather on power infrastructure. Examples of these sorts of situations are not hard to come by these days, and though they still represent a departure from ordinary life they should concern future-focused industry leaders.

The two major constraints that may shape the games of our climate changed future are likely to be the energy intensity and resource intensity of future games and devices. I put this challenge of designing within these sorts of constraints to game design students in the master's program at Abertay University in Sept 2019. One of the most productive discussions ended up around the question of how you would design a game if your player only had access to a computer for one hour a day. This scenario presented a significant shift in the expected norms of contemporary game design. Answers were offered along several lines—perhaps by reducing the importance of the computer in the role of the game so it is less essential, blending digital and analogue gameplay, or by incorporating lengthy preparation and planning in advance of access to computing resources. One suggestion involved designing a game in such a way that the player planned their decisions or actions and only input them for action once renewable power becomes available. This sort of game is hinted at by the Nintendo Gameboy remake known as the "Engage"—described as 'a video game console powered by a combination of energy from the sun and button-mashing during gameplay' (Ryan, 2020). Made through the collaboration of group of international engineering and computer science students, the device solves for the intermittency of solar resources by 'saving really, really quickly and restoring from our saved game really, really fast without anyone seeing' (Ryan, 2020). A similar theme emerged from our workshop, emphasizing the potential importance of asynchronicity to match the intermittent nature of certain types of renewable energy production. These might even mirror the dynamics of certain real-world experiences that games occasionally choose to reproduce. For example, giving orders on a battlefield and being unable to change those orders once given, mirrors the experience of a pre-modern general marshalling troops, of 'losing contact' with forces behind enemy lines, and generally being cut off from commands in real time. When we think about games that attempt to reproduce these kinds of situations—the *Total War* series of games, perhaps—they seem unlikely to go as far in that direction. There is always a risk that the consequences of this sort of design could very well be detrimental to the enjoyment of players, particularly given the way convenience and ubiquitous access to computation has conditioned the modern player to expect games to always be there whenever we want. Convenience itself has become such an intrinsic feature of much digital technology that the very idea of greatly energy constrained future seems unthinkable. But challenging this state of affairs might itself be generative of new possibilities. For now, games remain stubbornly wedded to the regime of fossil fuels and perpetual, unsustainable growth which we know will have to change. Whether our energy systems can adapt to meet our growing demand for energy (the supply side of the equation) or if they will need some forms of demand response as well remains to be seen. My bet is a combination of both, presenting an opportunity for some responsiveness and reductions in demand.

The history of computation also gives hints about the contours of contingent access to computation, when the earliest computers were shared, treated as a precious resource to be allocated and accessed judiciously. It is quite possible, depending on the political-economic pathways we choose to achieve energy grid decarbonization (or are forced into choosing, under certain grim scenarios) that the climate crisis could all too foreseeably force us to return to some sort of rationing of all but the most basic computational hardware, limiting personal energy use to when it is most available, perhaps reserving these resources for 'essential' work (the computationally intense work of climate modelling, perhaps). This is the extreme end of things, but it is impossible to rule out. The multiple times (first in 2017 and again in 2021) that high bitcoin and cryptocurrency prices have had powerful negative impacts on the availability of specialized silicon manufacturing, causing shortages of GPUs and even more general-purpose computing resources and pushing up prices perhaps points towards such a possibility, and the iniquities that comes with it.

Asynchronous play also has a long and grand tradition in games, from play-by-mail modes to other niche applications. The asynchronous game play of days-to-weeks long-term strategy games like *Neptune's Pride* (2010), where player orders are carried out slowly in real time could also become a solution, or a more prominent feature of game designs under such resource constraints placed upon us by climate change. In *Neptune's Pride*, players direct fleets of ships across a galactic map to conquer distant worlds, but crucially these fleets take hours or days to move, at what can feel like a glacial pace. Once orders are sent, they are unable to be rescinded (the fiction suggesting these fleets are locked into hyperdrive or similar). Part of the drama of the game is in the timing of these fleet movements, arriving at a system before an enemy attack. The joy and agony involved in striking before the enemy wakes up, or finding that while you were asleep someone has started an attack on your own system is much of the appeal. This leads to great anxiety over the timing of attacks, of retreats, of questioning alliances, and a hypervigilance at troop movements along borders. It is a complex interplay of social and game dynamics, as players plot and scheme behind the scenes. A certain degree of 'fog of war' is introduced simply via the time it takes between orders being sent and seeing their completion, as well as from the differences between players time zones, sleep patterns, and the rhythms of their play habits that bend (or clash) with the requirements of strategy (for dedicated players, at least). Stories abound from its heyday, of players setting up elaborate fleet movements in the middle of the night, setting alarms to wake and issue order at the precise moment their opponent is asleep, unable to respond. A reliance on intermittent renewables could structure games in a similar way, mandating completely different patterns of play that vary depending on players location, energy access, even seasonal variations in the availability of solar or wind. New challenges also appear under such a regime, reproducing or exacerbating existing inequities around energy access, or introducing them to parts of the world where they have not been at the forefront of designers' minds.

If graphical fidelity is a large determinant of real-time computation intensity (and therefore, of energy consumption) then perhaps a productive focus ought to be on shifting away from this. One of the observations that emerges from Mills et al.'s (2019) detailed analysis of the energy used during gameplay is the importance of real time metrics for energy and emissions intensity. They note that,

While gamers make intensive use of in-game diagnostics, energy use is not one of them. As instantaneous power feedback capabilities become available they should be effectively delivered to the gamer. Where enabled, developers may consider "gamifying" this information. Gamers seek out goal-driving systems for scoring and garnering merit for doing so. Carbon could be introduced as another variable. (Mills et al., 2019: 176)

This approach suggests leaving things to gamers themselves to respond to their energy intensity, and perhaps some might do so, but there is also potential to explore ways game designers and engineers might automate such a system. There are several APIs for the emissions intensity of national and regional energy grids that could be used to guide real time computational intensity of games, by scaling down graphics or other functions at times of peak emissions. The UK's Carbon Intensity API (National Grid ESO, 2021) allows for real-time measurement of emissions associated with a kWh of power at a highly granular level, allowing for this sort of dynamic adjustment. A similar service Watt Time (2020) is available for the United States that could be integrated, though it focusses more on optimizing the time of use of power by providing an estimate of when a region is experiencing peak emissions intensity. Similarly in Australia the national electricity market fuel mix is available in close to real time, allowing for the prioritization of energy use when renewables are at their peak of production (AEMO, 2021). Whether any of these techniques would be effective at producing substantial reductions in energy and emissions is a question that remains to be tested, and whether these sort of automations would be popular or acceptable to players is another. What it would reflect, however, is an ecological concern with the emissions that are a byproduct of playing the game.

Perhaps we can consider a field of more or less acceptable intrusions on gameplay that might achieve some of this same ecological attendance to a games own harms *in real time* (as simplistic as these initial suggestion may be). These might range from the drastic, like dynamic resolution scaling with a more 'pixelated' vision when the energy mix features few renewables, unlocking higher fidelity imagery when there is more; to locking framerates at 30FPS when emissions are high so that GPUs are artificially constrained; to enabling or disabling certain 'nice to have' graphics features when emissions go above or below certain thresholds. Many of these sorts of approaches might still prove to be little more than gimmicks, but it is my hope that professional developers will be able to come up with much more meaningful, and more effective solutions. Once again, the vision of a 100% renewable energy grid renders such constraints almost redundant, instead offering us a vision of plenty which I think most will find more appealing. This is important to emphasize, as it is already clear that climate ascetism has a limited appeal to anyone besides true believers.

Designers will likely find some of these suggestions underwhelming, unexciting, or even impossibly constraining from a technical standpoint. However, I think there is room to be optimistic even about constraints— one of the biggest challenges creatives face is the tyranny of the blank page. Game design as a practice has for many years gotten used to fewer constraints with each passing year, each new upgrade, each new console generation. It might stand to benefit from more constraints (even technical ones!) rather than less. Stories of some of the most innovative games in history often revolve around the limitations of early hardware. *Ars Technica's* "War Stories" series of interviews with designers of classic games from the 80s and 90s frequently reveals these constraints and lauds the creative solutions that led to some of the most loved innovations. Jordan Mechner, describing the development of the original *Prince of Persia* (1989) for the Apple II conveys a story of hitting the memory limit of the machine during development. The solution involved using a 'mirror' image of the player character as principal antagonist—an economical solution forced upon them by these external restrictions (Ars Technica, 2020b). Until the addition of an antagonist, Mechner claims, the game felt 'lacking' in many ways, and the idea of a magical 'mirror double' of the player became essential to the story. But without the memory constraints, which forced Mechner to rely on an economical "exclusive OR" (XOR) command of the Apple II to render the ghostly inverse of the player character's sprite, this essential part of *Prince of Persia*'s design is unlikely to have ever been realized. Dark mirrors of the player character have become a staple feature in many games since. Similarly, in an interview with Naughty Dog co-founder Andy Gavin (Ars Technica, 2020a), the original PlayStation console's limited memory and CPU speed necessitated a creative solution to streaming level and animation data from the spinning disc in real time. This even led to an early and important patent for Naughty Dog allowing them to license the technology to other developers, becoming an important part of their future business success. Dynamic streaming of game data and assets is now staple of modern game engines. Game development has a long history of flourishing inside of constraints, these sorts of limitations being responsible for some of its most memorable and important innovations. Perhaps an energy or emissions constrained future for games could be just as generative, and come to be embraced with gusto.

I have mentioned in passing in earlier chapters that energy constraints may also apply to the hardware side as, much like other computing devices, gaming hardware is extremely energy intense to produce—the few carbon neutral Xboxes mentioned in this Chap. 4 notwithstanding. Perhaps though, rather than offsetting the emissions of manufacturing, we should ask ourselves whether a better, more comprehensive answer is to make *existing* game hardware last longer, and slow down (even stop!) the current upgrade cycle business model. If the churning out of so much high-tech hardware is unsustainable (an issue which turn to in more detail in the next chapter), then hardware constraints will certainly be central to shaping the future games industry.

There is a persistent sort of fever-dream that often exists around in-development games, that they will be "the last game you ever play", in the sense that you will never need another, finally delivering on the full promise of what games can offer. What if we flip that on its head, and ask the rather more down to earth question of what if this is the *last console you'll ever need*? It is often remarked that the games that appear towards the end of a particular console's life cycle, when the hardware is much better understood, are often more able to fully exploit the known hardware, often seeing some of the most technically and critically lauded games. How much further could this be extended? What would experienced developers stand to gain, what new heights of technical proficiency and innovation could be achieved if we did "more with less" instead of automatically upgrading to the newest version of this or that engine or the latest piece of hardware? Is it possible to interrupt the graphical arms race? What would it take?

I think of the games that arrived at the end of the original Xbox era and what they were able to achieve within the constraints of the ageing hardware system. Particularly a game like *Halo 2* which I remember as a technical and artistic tour de force, squeezing unbelievable environments and effects from the Xbox, reflecting a profound understanding of the technical strengths and restrictions of the system. As someone who had an Xbox 360 in their household not long after launch, I distinctly remember *Halo 2* remaining the best game on the next generation of consoles for a long time in my eyes, a testament to just how much of an achievement it was, and having little to do with what CPU or GPU it was running on.

What if gaming after climate change *necessarily* means giving up the energy and resource intensity of the hardware upgrade cycle? Can we imagine a situation where this cycle (having already slowed down dramatically) is elongated further still, perhaps even to the point at which new hardware and new gaming devices become all but redundant? What if legislation, or some other dramatic climactic necessity intervenes for us? What if we faced this generation as *the last console generation*, with consumers, developers, and climate leaders saying instead of "what are the specs of the next PlayStation going to be?" rather wondering "who is going to do something revolutionary with what we've already got?" What sort of games could be made if the industry came to its senses about the diminishing returns of pixel pushing?

If console and gaming hardware upgrades cease, at the very least games will need to find other criteria on which to market and differentiate new games, as the scope for technical advancements such as higher resolutions, greater visual fidelity, and so on, become markedly restrained. The growing trend towards games as a service—from *Fortnite*'s Battle Passes, *Apex Legends*' live events, and *Destiny 2*'s seasons revealing new story beats weekly—new ways of keeping players engaged and interested that don't require buying brand new games and brand-new hardware sold to them with the false promise of revolutionary new graphics become much more plausible. Perhaps it could even shift the industry's emphasis onto novelty and innovation through mechanics, design, style, artistry and storytelling, over sheer graphical grunt, effusive new shader effects, and the endless proliferation of particle systems. We can see prototypes for this tendency in cultures adjacent to games already, though in some unlikely locations. From early on, games have been closely related to the so-called "Demoscene"—with studios like DICE, which produces the *Battlefield* series of games, founded by former members of the demoscene. This unique subculture represents a hyper-prioritization of efficiency in code, valorizing technical skill, and above all the ability to "do more with less" (Borzyskowski, 1996; Garda & Grabarczyk, 2021)—all features that may prove essential to the carbon neutral games industry of the future.

Similarly, Nintendo has for much of its history made great use of what some consider "underpowered" or "last generation" hardware, combined with off-the-shelf parts. Eschewing bespoke hardware even as it focused on alternative interfaces it also has a habit of designing and marketing games along different lines than who can push the most pixels to the screen the fastest. From Nintendo's portable devices that dominated the handheld gaming market before and after the rise of mobile gaming, eschewing the power of handhelds like the PlayStation Portable, to the remarkably innovative but underpowered Wii, described by some more tongue-in-cheek commentators as "two game cubes strapped together", this device still managed the kind of popular appeal that saw what Juul described as a 'casual revolution' in the games industry (Juul, 2010). Nintendo has benefited greatly from opting out of the graphical arms race, and shown repeatedly that it is possible.

In the case of Nintendo consoles, however, it must be noted that the disparity in hardware performance as a result represents a real, substantive difference in computational capacity. It means fewer numbers of physical transistors, fewer discrete metal-oxide semiconductors for storing bits, and fewer graphical processing units to rapidly do the maths of moving triangles and millions of different pixel colours to output to the screen somewhere on the order of thirty-to-sixty times per second. But this is precisely the goal, as each of those transistors comes with a cost to the planet. As a result, however, there are simply some things that a Nintendo console cannot do as a result of its much lower computational capacities. We should also understand that there are real pleasures to high powered graphical fidelity—but that it is equally impossible to disentangle the desire for these from their production by the game marketing apparatus, which devotes millions annually to proving the superior nature of new games and game technologies. However, rather than seeing the limitations of Nintendo devices as some sort of "failure" (as they can tend to be seen by hyper-partisan consumers) these limitations can become their own unique selling point. Nintendo has positioned itself as a different kind of console maker and devices like the Wii, the 3DS, and the Switch as a different environment for a different kind of game, selling them with a promise that emphasizes style and design. What prevents the games industry from doing this on a wider scale? Can the barriers to this be removed in time to slow down and pull us back from the cliff we face in the headlong rush to ever more energy intense, ever more computationally demanding games? Perhaps the question to ask here is again, where, how, and why are the desires for certain types of games made? These are uncomfortable but necessary questions to ask. How much has the games industry cultivated this desire itself, and how much can be said to be 'innate' in the player themselves? Graeme Kirkpatrick's (2015) analysis of the construction of the 'gamer' identity and culture through the formative period of 80s and 90s leads to some uncomfortable conclusions about the balance of responsibility.

Most of these examples in this outline of possibilities have largely taken ideas from the pool of already-existing games, and looked at where some might not fit the constraints we face in the future. This might tend to suggest that the current argument is about closing down game design possibilities in the time of climate change. But I do not think this is really the case—it can't be proven, but experience suggests that our track record on predicting the future, and particularly the results of major paradigm shifts has often been exceptionally poor. Game designers may very well respond to these and other, unexpected new constraints with a burst of stunning creativity along lines completely different from these, or which are radically unthinkable from the perspective of the now. That is my hope, and in this way I find thinking about possible games after climate change a productive and positive exercise today. Doing so *now* can only encourage the development of the sort of creativity we will need much more of in the years ahead, a creative blossoming that will be needed to take us in all sorts of unexpected new directions.

I am reminded of David Kanaga's wonderfully eclectic *Oikospiel Book I: A Dog Opera in Five Acts* which features a 'box office' store front with a ticket price that reduces as the buyer moves their mouse around on the screen, powering a virtual windfarm generating a miniscule discount (1c at a time!). The price of the game itself reduces to a minimum of $5 as the player "powers" the website with their hand motions. Despite not being connected to a real windfarm the image is an evocative one—drawing attention to material existence via enrolling the player in a time/cost calculation (how long are you willing to sit there and wave your hand around?) and the electrical power required for it. It also features as far as I know the only purchasing system with a slider for "household income", and accounting for number of dependents (if you make over a 1.4 million dollars a year, Kanaga would like to price the game at $14,000 for you—which seems very fair). This focus on equity in pricing is deeply surprising and impressive, doing far more to account for the variety of subject-positions of the player that I called for in Chap. 2, particularly their socioeconomic class, better than any other that I know of. These decisions are made all the more special by the fact that Kanaga is foregoing real money to make this point.

We could also project game making into a better future scenario—this time, a world of plenty, in which the international community is entirely successful at curtailing carbon emissions, a scenario that would feature plentiful even excess renewable energy, and removing carbon emissions entirely from power systems the world over. However depending on how this is achieved, such as through the imposition of price mechanisms or through an equitable and just transition, we might still end up back in the situation precisely that I had in mind at the introduction to this book. There I proposed that the day might come that sees players choosing between the air-conditioning and their favourite games, and this cannot be separated from consideration of global climate justice (much like in Kanaga's synthesis above). For developing parts of the world, electricity supplies can already be intermittent or unreliable, and improving reliability for millions even as the climate crisis deepens and becomes more disruptive is a challenging task. What effect would this have on the class makeup of game players? Would gaming become the preserve of the global rich even more so than at present? This could have an impact on the markets that exist for certain kinds of games, and ironically the best adapted segments of the gaming industry are *not* likely to be Triple-A gaming, but the mobile gaming parts of the industry—with devices that can be charged whenever power becomes available and used even when grids are unstable, renewables are not generating enough, or offline. The advantages that mobile phones have as a platform for gaming in a time of climate change are already substantial, with drastically lower power consumption. Mobile gaming is not without its downsides, of course, many of which are already evident: proprietary platforms controlled by Apple dominate the paid games marketplace and a much disliked F2P ecosystem exists across both Apple and Android, among other downsides like relatively limited input methods. Mobile gaming represents simply too much of a compromise for some. But as the lowest energy intensity platform, games as a whole could stand to learn much from them.

There are also just ordinary business and technical challenges to prioritizing energy efficiency, one particularly illustrative example being the *Warcraft 3 Reforged* remake released in January 2020. As discovered and disseminated by twitter user Colin Carnaby (2020) in a series of tweets, the game uses a highly inefficient Chrome web browser for its main menu interface, using far more computational resources in the menu than it does during actual gameplay. Why is this the case? Chrome was originally designed to be a secure and powerful browser, but it has become such a computationally demanding application that it becomes unstable (it repeatedly crashed the nearly new iMac that I used to write parts of this book, forcing me to change to a different browser). Chrome seems an awkward fit for this particular use, but as a quick solution it makes sense in the rushed circumstances of game development. Perhaps business executives, publishing deadlines, or even UX design experts demanded an up-to-date launch interface experience based on user feedback.

I want to emphasize that I am in no sense a game designer, and many of the actual design implementations which could arise in an energy constrained future would be radically different (and radically more inventive) than these very initial and untested sketches. Just like many readers, who might justifiably balk at the prospect of some of the above scenarios and their solutions, I too would prefer an entirely emissions free future, filled with plentiful power and where many of these types of concerns become moot. But until we are guaranteed such a future, the benefits of planning for these constraints, and doing what we can to reduce emission today should be clear.

## CONCLUSION

In this chapter we looked at the emissions of game playing, with a combination of measured and estimated energy figures for specific consoles, and aggregate emissions for large groups of players, informed by Aslan's (2020) extraordinarily detailed estimates of PS4 emissions in Europe, and Space Ape Games' clear and comprehensive methodology for game developers to measure and offset Scope 3 emissions from their players. The economic feasibility of this sort of offsetting across the industry remains an open question, representing a carbon cost in the hundreds of millions for the whole sector. For now, the best placed to pay those costs are those making games for the most power-efficient gaming devices (like smartphones) precisely because these are the least emissions intensive devices, and the costs remain manageable. The largest portion of player emissions from console and PC gameplay remains a far more expensive and formidable offsetting prospect. I eagerly await the first Triple A game studio or publisher who is able to, at a minimum, report emissions figures for their players, and then somehow come up with the money to offset the climate impacts of player activity. This is a necessary pre-requisite to the ecological game, acknowledging the impacts of its own existence on the wider world.

The goal of this chapter has been to try and provide informative energy use figures for the most popular gaming devices to get a sense of the scale of emissions that come from playing games now and into the future. I have not attempted an estimate of all gameplay emissions from play, as this is extremely complicated, though we have seen estimates from Mills et al.

(2019) for emissions in the United States in 2016 of around **24 million tonnes of $CO^2$-equivalent**, and Aslan's (2020) figures for the total energy use of the PS4 install base from launch up to 2020 reaching as much as **18.6 million tonnes of $CO^2$-equivalent** emissions. Both of these figures suggest end-user emissions that dwarf the industry's own scope 1 and 2 emissions, estimated in Chap. 4 at **0.5–15 million tonnes of $CO^2$-equivalent** emissions for 2020. What is less important, however, than exact figures for today or even tomorrow, much as I emphasizes in Chap. 5, is the knowledge that emissions exist in the first place, having a sense of where they come from, and also some ideas about what might be done to abate them. It is important to remember that end-user emissions, despite large potential avoided emissions even from small reductions, should come *after* the work of reducing the Scope 1 and 2 emissions, which are the more immediate responsibility of the games industry. There is almost no downside to reducing emissions in these areas, it can be done quickly, cheaply, and may have an outsized impact elsewhere by accelerating the renewable energy transition, particularly when large corporations use their capital to enable investment in new renewable power generation. Similarly, the emissions arising from shipping discs around the world already have plausible and attractive alternatives, despite some downsides such as less freedom and control over reselling games, making the full-digitization of games distribution more controversial a prospect. The emissions of end users and the reduction of energy demand through game design, by contrast, while potentially necessary in some scenarios also comes with significant downsides for end users. Mills et al. (2019) acknowledge this in their analysis by focusing on energy efficiency initiatives in devices as the best way to achieve gains without having to sacrifice things many users find pleasurable. Some players—and I readily admit to being among them at times—take great pleasure from cutting-edge graphics technology, and we cannot treat this as solely the product of misguided or ideological views (even as the industry's role in deploying it as a marketing tool in the past has surely contributed to it as a present demand). As I argued in Chap. 2, the creation of these desires—the context that cultivates and brings them forth in players, that reproduces the photorealistic industrial complex and an audience for products—should be the target of our interventions. But that is a big ask, all the same. So too is the ask for no new consoles and hardware, to reprioritize to doing more with what we already have instead of adding to future landfill and current resource extraction processes required for making devices. We turn to this topic in more detail in the next chapter.

The three major areas of emissions from the games industry that Chaps. 4, 5, and 6 have presented, I argue, deserve addressing in roughly that order, starting with the highest priority and easiest reductions to be made. Starting with the emissions intensity of making games, focusing on getting the games into the hands of players in a carbon neutral way, and finally figuring out ways to bring down emissions from players playing games—all are useful and necessary. These chapters have shown that the games industry will have to reduce the emissions it's responsible for as well as the resources and energy demand that it helps create if it is to achieve carbon neutrality. It will need a strategy for achieving these changes—they will not happen automatically, and there may well be resistance from some quarters. But as long as demand for new devices persists, reducing the emissions and other environmental harms from supply chains and other extractive processes that underpin modern high-tech manufacturing will be necessary. This is the final source of emissions to discuss, one which also broadens our attention *ecologically* to consider the other kinds of harms entailed beyond carbon emissions. Device manufacturing as a source of emissions might even prove the most tenacious, hardest to address aspect of the contemporary games industry. This final step in the path to decarbonization entails substantial challenges, distributed across the entire globe, and involves a host of environmental costs baked into consoles before we even boot them up for the very first time.

## REFERENCES

Apple. (2017, September). *iPhone 7 Plus Environmental Report.* https://www.apple.com/environment/pdf/products/iphone/iPhone_7_Plus_PER_sept2017.pdf

Ars Technica. (2020a). How Crash Bandicoot Hacked The Original PlayStation | War Stories | Ars Technica. *YouTube.* https://youtu.be/izxXGuVL21o

Ars Technica. (2020b). How Prince of Persia Defeated Apple II's Memory Limitations | War Stories | Ars Technica. *YouTube.* https://youtu.be/sw0VfmXKq54

Aslan, J. (2020). *Climate change implications of gaming products and services.* PhD dissertation, University of Surrey.

Australian Energy Market Operator (AEMO). (2021). *Australian National Energy Market Dashboard.* https://aemo.com.au/en/energy-systems/electricity/national-electricity-market-nem/data-nem/data-dashboard-nem. Accessed 30 Mar 2021.

Borzyskowski, G. (1996). The hacker demo scene and it's cultural artefacts. *Cybermind Conference, 1996*, 1–23.

Carnaby, C. (2020, January 31). I've been trying to figure out why the main menu performance on Warcraft 3 Reforged is so bad. It ends up the whole main menu is a web app running on Chrome. This thing runs worse than the actual game (likely because it's pegging an entire core of my CPU). *Twitter.com*. https://twitter.com/colincornaby/status/1223073101312753664

Claburn, T. (2021, July 26). Dell won't ship energy-hungry PCs to California and five other US states due to power regulations. *The Register*. https://www.theregister.com/2021/07/26/dell_energy_pcs/. Accessed 7 Sept 2021.

Delforge, P., & Horowitz, N. (2014). The latest-generation video game consoles. *Natural Resources Defense Council Issue Paper IP*, 14–04.

European Commission. (2021, July 14). *Carbon border adjustment mechanism: Questions and answers*. https://ec.europa.eu/commission/presscorner/detail/en/qanda_21_3661

Frumusanu, A. (2019, October 16). The Apple iPhone 11, 11 Pro & 11 Pro Max review: Performance, battery, & camera elevated. *Anand Tech*. https://www.anandtech.com/show/14892/the-apple-iphone-11-pro-and-max-review/4. Accessed 1 Apr 2021.

Garda, M. B., & Grabarczyk, P. (2021). "The Last Cassette" and the local chronology of 8-bit video games in Poland. In *Game history and the local* (pp. 37–55). Palgrave Macmillan.

Juul, J. (2010). *A casual revolution: Reinventing video games and their players*. MIT Press.

Kim, M., Kim, Y. G., Chung, S. W., & Kim, C. H. (2014). Measuring variance between smartphone energy consumption and battery life. *Computer, 47*(7), 59–65. https://doi.org/10.1109/MC.2013.293

Kirkpatrick, G. (2015). *The formation of gaming culture: UK gaming magazines, 1981–1995*. Springer.

Malik, M. Y. (2013). Power consumption analysis of a modern smartphone. *arXiv preprint* arXiv:1212.1896.

Mills, E., Bourassa, N., Rainer, L., Mai, J., Shehabi, A., & Mills, N. (2019). Toward greener gaming: Estimating national energy use and energy efficiency potential. *The Computer Games Journal, 8*(3), 157–178.

National Grid ESO. (2021). *Carbon Intensity API*. https://carbonintensity.org.uk/. Accessed 30 Mar 2021.

Rovio. (2020, March 6). *Rovio entertainment corporation corporate responsibility report 2019*. https://investors.rovio.com/en/corporate-social-responsibility

Ryan, J. (2020, September 2). The first battery-free Game Boy wants to power a gaming revolution. *CNet*. https://www.cnet.com/features/the-first-battery-free-game-boy-wants-to-power-a-gaming-revolution/

Sorrell, S. (2009). Jevons' Paradox revisited: The evidence for backfire from improved energy efficiency. *Energy Policy, 37*(4), 1456–1469.

Space Ape Games. (2020). *We've gone carbon neutral: Here's why we did it, and how you can too.* https://spaceapegames.com/green. Accessed 30 Mar 2020.

Watt Time. (2020). *Watt Time – The power to choose clean energy.* https://www.watttime.org/. Accessed 30 Mar 2021.

World Energy Council. (2016). *World energy scenarios: The grand transition.*

GAMES

Blizzard Entertainment, Lemon Sky Studios. (2020). *Warcraft III: Reforged.* Blizzard Entertainment.

Broderbund. (1989). *Prince of Persia.* Broderbund.

Bungie. (2017). *Destiny 2.* Bungie.

Bungie. (2001). *Halo: Combat Evolved.* Microsoft Game Studios.

Bungie. (2004). *Halo 2.* Microsoft Game Studios.

Creative Assembly, Feral Interactive. (2000–). *Total War* series. Electronic Arts, Activison, Sega.

David Kanaga. (2017). *Oikospiel Book I: A Dog Opera in Five Acts.* Itch.io.

EA DICE. (2002–). *Battlefield* series. Electronic Arts.

Epic Games. (2017). *Fortnite.* Epic Games.

Iron Helmet Games. (2010). *Neptune's Pride.* Iron Helmet Games.

Respawn Entertainment. (2019). *Apex Legends.* Electronic Arts.

# CHAPTER 7

# The Periodic Table of Torture

A great deal of work over the past decade or more by scholars and journalists has increased our awareness of the harms involved in the ever-
increasing streams of e-waste, produced in staggering quantities by the
explosion of digital media technologies and the churn of new devices
(Grossman, 2006; Maxwell & Miller, 2012; Taffel, 2012, 2015; Cubitt,
2016; Fuchs, 2008; Rao et al., 2017; Oguchi et al., 2011). This important focus on e-waste and the end-of-life of devices, however, can at times
have the unfortunate side-effect of overshadowing the harms that have
*already* occurred well before it is even bought. These are harms that are
specific to a particular device, its material composition and manufacturing
processes, but also occurring across the entire category of digital devices,
from smartphones to desktops and consoles. Substantial attention, again
quite rightly, has been directed at the harms caused by the mining and
sale of minerals that facilitate wars and conflict, like coltan in the
Democratic Republic of the Congo, (Nest, 2011) along with others
members of the group of so-called 'conflict-minerals'—tin, tungsten, tantalum, and gold (two of which appear in this chapter's list of elements).
Materialist media studies researchers and environmental journalists focusing on the materiality of devices have done much to raise awareness of the
chemical compounds, heavy metals, plastics and other toxic or carcinogenic substances present in digital devices and the problems they create
for waste management, lingering in the environment with serious

© The Author(s), under exclusive license to Springer Nature        179
Switzerland AG 2022
B. J. Abraham, *Digital Games After Climate Change*, Palgrave
Studies in Media and Environmental Communication,
https://doi.org/10.1007/978-3-030-91705-0_7

consequences for years (Grossman, 2006; Oguchi et al., 2011; Rao et al., 2017). Yet in spite of this growing awareness it still remains difficult to draw clear, conclusive connections from a given device to the *particular proven harms* it has caused. Primarily we have been limited to a focus on categories of materials and a broad awareness of the problems associated with resource extraction. While still important, these arguments can end up disconnected from the actual purpose or function of these materials, leaving researchers, policymakers and other decision takers ill-equipped to know why these materials were used in the first place, whether they were necessary, let alone what can be done about them when they enter waste streams as millions of discarded electronic devices. Most descriptions of the material composition of devices remains extremely high-level and conceptual, rarely approaching the specifics of hardware function. Without reference to this kind of detail, we are left relatively disarmed in the face of the tacit assumptions of industry and consumers apparently ravenous appetite for newer, faster, better devices.

In this chapter I tell a series of connected stories about the environmental burden of manufacturing the PlayStation 4 APU (standing for the 'advanced processing unit'—a single chip that combines both CPU and GPU functions). I do so by following its elementary constituents back to their origins and the processes involved in digging them out of the earth's crust. For each of the seventeen different atomic elements detected in the chip in substantial quantities, I outline connections between the metals in the device, the environmental harms associated with its mining, refining, transport, and manufacture. Most importantly of all, I attempt some educated guesses and informed speculation in the search for explanations about why each of these metals and metalloids might be present, and what they are doing in the device. Friedrich Kittler's (1995) provocative essay 'There is no software' makes a strong argument for attending to hardware—circuits, logic gates, transistors and other components under the hood of digital devices—as the essential level of computation. His provocation suggests that 'all code operations… come down to absolutely local string manipulations and that is, I am afraid, to signifiers of voltage differences.' Without understanding the material substrates that code operates on we cannot possibly hope to understand computation (and by implication, gaming). Attending to the very stuff that enables, regulates and directs these voltage differences—the signals and voltages that fundamentally all digital games cannot exist without—is critically important for our understanding of not just what these machines do, but what sort of consequences are entailed for the material world itself through our choice to make these sorts of devices, and to play games with them. Specific details like these have been sorely missing from even the materialist study of games, allowing for the environmental and climactic burdens of game hardware to fly under the radar.

Grossman (2006: 20) notes that our lack of knowledge in this area is partly due to the secretive nature of the highly competitive business of semiconductor manufacturing:

> While the categories of materials are fairly consistent, it's hard to get a precise list of ingredients involved in the entire process of manufacturing a semiconductor or circuit board. The specific recipes for these products are generally proprietary, and materials vary depending on the kind of microchip being produced and its fabrication process. What's more, materials change continually as new products are developed.

This chapter offers the kind detail that Grossman (2006) notes is so often hard to get, providing a precise list of the atomic elements that constitute the nerve-center of the *PlayStation 4*—a staggeringly complex array of materials in the small package that is the APU. The discussion of each element in the list allows for new an entirely new level of detail and far greater confidence in our ability to attach specific harms—from $CO_2$ to poisoned waterways—to this device that make games possible. Ultimately, the goal is to inform future decisions about the resource intensity of these devices, the manufacturing processes involved, and ultimately whether or not we are willing to accept these harms. I hope that reading about them helps motivate those involved in perpetuating the demand for new devices to stop, acknowledge the seriousness of what is entailed so casually and automatically, and perhaps consider alternatives. By knowing more precisely what might be entailed in, say, a new generation of PlayStation consoles, we might have a better idea of whether or not it is worth the cost, and what might be done differently.

The progenitors of the type of work I am doing in this chapter lies in a number of directions. Methodologically, this chapter was first inspired by the project of technology journalist and author Brian Merchant, (2017) who with the assistance of a commercial metallurgical testing company, performed the same kind of analysis on an iPhone, with the goal of discovering what the device is made out of at an atomic level. In his case this involved crushing and grinding a whole iPhone into a fine powder, capturing the escaping gasses and other volatile compounds, and analyzing the spectrographic profile of the homogenized particles. The test provided important insights into a particular class of digital object that has become a ubiquitous focus of materialist media studies as well as an object of great consumer interest. Crucially it provided data about a *particular* iPhone model—the iPhone 6 16GB circa 2014, to be precise. Digital devices, despite common features, are not identically composed. Smartphones in particular have been the object of intense and sustained engineering refinement for over a decade, with regular annual releases of upgraded and improved components and hardware. Other devices receive somewhat less intense engineering attention, yet as we saw in Aslan's (2020) discussion of the power consumption improvements in PS4 models over its lifetime, gaming devices also change across hardware iterations with substantial energy efficiency improvements possible.

Some of the earliest work that informs this chapter comes from environmental journalists, like Elizabeth Grossman (2006) who raised early alarms over a raft of issues around the waste products of the glittering high-tech industry. As research for her book *High Tech Trash*, she travelled to copper mines and other sites of extraction of the resources that are indispensable to semiconductor manufacturing. These are the often hidden yet-essential parts the modern world. Grossman's work features in Rayford Guins (2014: 5) book *Game After* about the post-consumer life of games, who also explains why this sort of work is so rare:

> In the field of game studies and within the consumer market we rarely (or are only beginning to) attend to video games... in terms of their aging, deterioration, obsolescence, ruinous remains, or even history.

The history of each of the elements discussed, and the impacts each one has had on various parts of the world is indispensable to our understanding of where these devices have come from, and what they have cost us. Similarly, Taffel (2012: 1) argues the physical hardware and devices used in the digital media industries create 'material social and environmental justice issues which have thus far received scant attention within media and communication studies.' Benjamin Bratton's (2016) *The Stack* does important theoretical work in this area, offering a series of connected geological, and material-theoretical frameworks for considering the internet, the spread of digital devices, platforms, and communications technologies as an 'accidental mega-structure... a new architecture for how we divide

up the world into sovereign spaces' (Bratton, 2016: xviii). Bratton point-
edly directs our attention towards the mineralogical, noting that 'there is
no planetary-scale computation without a planet, and no computational
infrastructure without the transformation of matter into energy and energy
into information' (Bratton, 2016: 75). Planetary scale attention and interventions are now critical, as we see underscored by Andrae and Edler's
(2015: 144) shocking projections for the potential energy consumption of
the internet and the world's growing communications infrastructure—
worst case scenarios suggesting it could be responsible for as much as half
of the world's electricity use or 'up to 23% of the globally released greenhouse gas emissions in 2030.' Bratton (2016) describes the geological
vectors produced by the internet and high-tech industry, working overtime at transforming the earth into an increasingly power-hungry infrastructure of control and surveillance (Bratton, 2016: 76). Taffel (2015:
80) importantly reminds us, however, that as tempting as it is to view
e-waste arising from digital media technology as 'a historically novel situation, the externalization of deleterious health impacts onto workers associated with technological production and disposal is a phenomenon with
links back (at least) as far as the industrial revolution.' While many of the
upsides of the modern digital media industries are new, a lot of the downsides are very, very old.

In an analysis of e-waste streams as a potential resource for important
metals, Oguchi et al. (2011: 2155–6) indicated that 'video games have the
highest potential as secondary resources of precious metals. ...video games
should be considered as a target to be exploited as secondary resources of
precious metals.' The global install base of PlayStation 4's in 2020 was
about 100 million devices—how much of these precious metals are on the
cusp of being thrown away right now? What materials are in these devices
that, which are thrown out with almost shocking regularity? What sorts of
planetary resources is the perpetual upgrade cycle using up and what can
be done about it? Where do the materials for this device come from and
what did it take to transform them from raw materials to useful computational infrastructure? The periodic table of torture begins to answer some
of these questions.

Other researchers have looked at the structural elements of the PS4 and
its environmental impact as well. Lewis Gordon (2019) a journalist with
The Verge, obtaining the assistance of a Cambridge University materials
laboratory, helpfully provides details about the composition of the plastic
and metal shell of the PS4. He describes the researcher's findings as they

disassemble the console, finding the majority of the outer black plastic to be acrylonitrile butadiene styrene (or 'ABS') probably the most common form of molded plastic in use. They note that 'an origin is engraved into the plastic: the Casetek Computer factory in Suzhou' (Gordon, 2019). This, as well as a mark labelling it as ABS, 'means it has a better-than-normal chance of being recycled' (Gordon, 2019). The laboratory weighed 511 grams of ABS in the console's case, putting a figure of 'approximately 1.6 kilograms of carbon dioxide equivalent into the atmosphere for every PlayStation 4 manufactured' (Gordon, 2019). A further 736 grams of steel is measured, with Gordon noting 'the proportion of recycled material in new steel sits at around 35 percent. In China, where the PlayStation 4 is made, it's even less, clocking in at 20 percent.' According to Norgate et al. (2007) who provide emissions figures for many common metals, producing just this much steel generates about 1.7 kg of $CO^2$ equivalent emissions. An aluminium heatsink, copper heat pipes, and a variety of ceramics and wound copper wires inside the power supply, further adding to the emissions bill prior to assembly.

The result of the analysis that Gordon and the Cambridge research lab that undertook put an approximate figure on the $CO^2$ emissions of a single PS4: 'The equivalent of 89 kilograms of carbon dioxide is emitted into the atmosphere with the production and transportation of every PlayStation 4.' This is a staggering figure, but by no means extraordinary in terms of high-tech devices—in the same piece, the head of the Cambridge research laboratory Claire Barlow notes that the figure is not significantly more or less than any other device. At their heart, every computational device has something that does the heavy-lifting—the logical work that gives it the power to be programmed, to do what we ask it to at speeds far greater than the human mind can comprehend. In the PlayStation 4 this is the APU.

## Testing the Advanced Processing Unit

The focus of this chapter is the beating heart of the PS4 itself—the combined CPU and GPU in a single system-on-a-chip that does almost all of the fundamental, basic computational maths inside the PS4. The combined CPU and GPU chip was designed by tech giant AMD and produced by semiconductor company TSMC and it is almost guaranteed to be the single most energy intense part of the console. Although it is based on a now quite old 28 nm process (according to ChipWorks a reverse engineering outfit) compared to the PS5 and its state-of-the-art 7 nm process, it represented something fairly close to top-of-the-range on its release in 2013. The APU itself has a relatively low clock speed—only 1.6 GHz—compared to the more common 3+ GHz of 2020's multi-core Cpus. Clock speeds are not everything, however, and the architecture of console chips and their standardized hardware configuration provides other benefits to software developers, often allowing them to maximize the potential of the hardware. As the 'brains' of the console, the chip does both logical operations of the CPU, as well as all the high-speed parallel processing required for real-time 3D graphics. The APU chip is what we might consider the most 'essential' part of the console—the part that makes it what it is and gives it its unique capacities. As the nexus of so much scientific and engineering development, the cause of a great expenditure of resources, it is the focus of our current investigation (Fig. 7.1).

Having had a number of iterations over its lifespan, the PSDevWiki lists several models of APU. The particular PS4 console that I purchased via eBay for $80 was almost certainly one of the first four APU

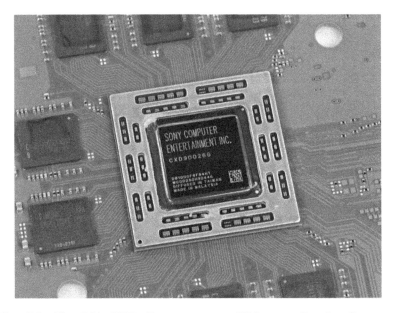

**Fig. 7.1**  The PS4 APU. (Image courtesy iFixit.com, Creative Commons BY-NC-SA Walter Galan. Used with permission)

models—either the 26G, 26AG, 37G or 43GB. Unfortunately, I neglected to record the model number before it was subjected to the destructive testing, however we can narrow it down at least one of the four chips—all produced prior to the PS4 slim model that replaced it in September 2016. The console was listed by the seller I bought it from as broken, 'for parts only', and the APU unaffected by the other damage on the board. Many of these APUs can still be bought online through shops like AliExpress, with prices around the $45 USD mark in late 2020, however, due to the complexity of their connection to the motherboard they are rarely if ever replaced. This is down to the use of a 'ball-grid array' method of soldering the chip to the board, rather than the more complicated, but more user-replaceable 'socket' of most consumer PCs, which allows the swapping in and out of CPUs with relative ease. Using a ball-grid array to connect the CPU to the main logic board reduces manufacturing complexity and costs, and the overall weight slightly, but it results in a much harder-to-repair device. Even dedicated computer repair shops are unlikely to have the necessary equipment to make replacing the APU an option, despite being a widely available part. It is simply often cheaper to replace the entire board (Fig. 7.2).

In the sections of the chapter that follow I offer a map, not of where, but of *what* atomic elements makes up this centrally important

**Fig. 7.2**  Full teardown of the PS4 console. (Image courtesy iFixit.com, Creative Commons BY-NC-SA Sam Goldheart. Used with permission)

component. The chapter is organized around the elements of the periodic table, focusing on those detected inside of a small (less than half a gram) sample of this second hand, broken PS4 that I purchased for the purpose. To come up with the list, I had the University of Technology, Sydney science laboratory conduct an inductively coupled plasma mass spectrometry test (ICP-MS) test similar to what Merchant (2017) did to an iPhone. The sample was to be dissolved in a strong acid, having first provided the lab team with research papers demonstrating a safe method for the test to be conducted on similar components. This was necessary as damage to the ICP-MS equipment can occur if the test material is not entirely dissolved, and this would prove costly (on the order of tens of thousands of dollars). Once satisfied of the proper procedure, a tiny section of the APU was separated from the rest of the die via powerful guillotine, lab technician Dayanne Mozaner Bordin dissolved this sample in a combination of 4 mL of analytical-reagent grade nitric acid and 8 mL of analytical-reagent grade hydrochloric acid, placing the entire solution in a hot water bath at 90 °C for approximately 48 hours until the sample was fully digested. This mixture was then cooled to room temperature, 50 mL of MilliQ ultra-pure water was added, and the sample processed using ICP-MS, with small amounts of the dissolved liquid passing through a high powered stream of plasma in a tightly controlled environment, resulting in its separation out into various atomic particles that were then measured and identified using mass spectrometry. Three 'blank' samples were included as part of the test process, to ensure detection was genuinely from the presence of elements in the sample rather than cross contamination, and a standard mixture containing all the elements to be tested for was used as a control. The ICP-MS test of the sample was conducted in triplicate, verifying the results and the presence of the identified elements. The meticulousness of the above process means we can be very confident of the findings, and of the presence of the following atomic elements. The test was not calibrated for measuring exact quantities (a much harder and more expensive task) however the test was able to identify certain elements that are *likely to be present in greater quantities*, and it is on these elements that this chapter focuses. At some point in time, somewhere in the world, each of these elements have been dug out of the ground, refined, transported, and transformed into the brains of this game console. There are 54 different elements detected, having tested for at least 78 different isotopes (each element can appear in slightly different forms or 'isotopes') and these are displayed on the period table in Fig. 7.3. below. The elements of the periodic table

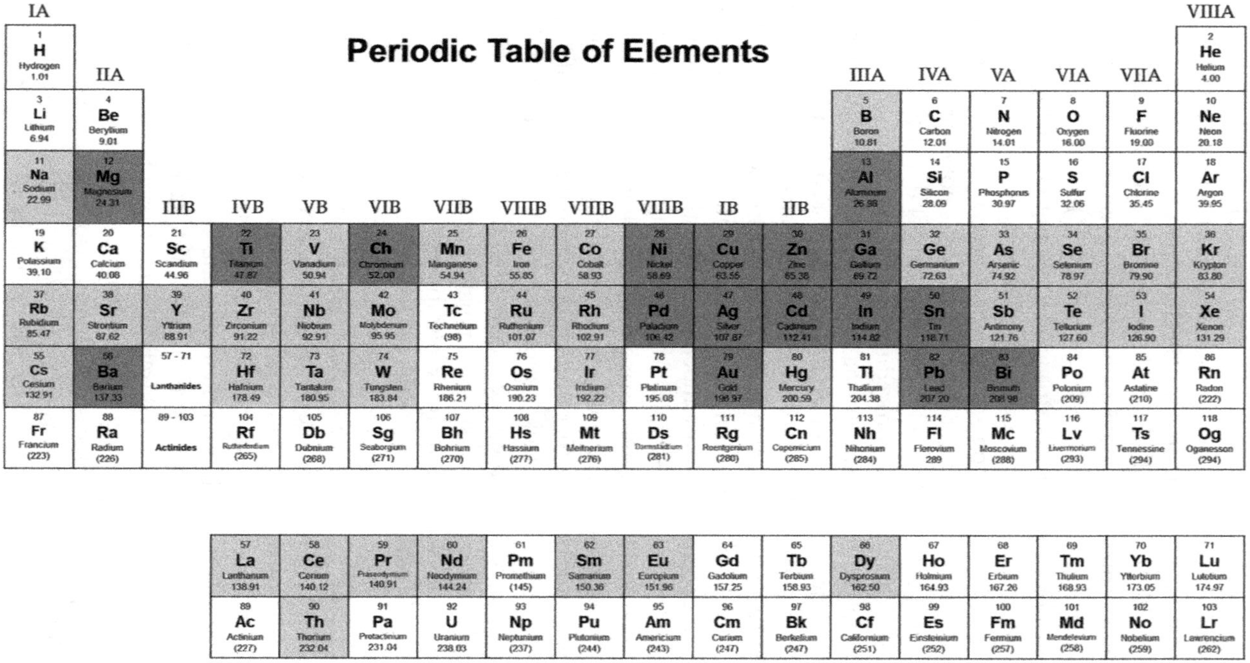

**Fig. 7.3.**   CC BY-SA 4.0—Emeka Udenze, with additional shading to indicate elements present in the sample. Light grey indicates detection in the sample, dark grey indicates detection, likely in higher quantities

shaded in both light and dark grey were identified, but only those in dark grey were likely to be present in higher concentrations.

Listed according to their atomic order, the elements detected were: boron, sodium, magnesium, aluminium, titanium, vanadium, chromium, manganese, iron, cobalt, nickel, copper, zinc, gallium, germanium, arsenic, selenium, bromine, krypton, rubidium, strontium, yttrium, zirconium, niobium, molybdenum, ruthenium, rhodium, palladium, silver, cadmium, indium, tin, antinomy, tellurium, iodine, xenon, cesium, barium, lanthanum, cerium, praseodymium, neodymium, samarium, europium, dysprosium, hafnium, tantalum, tungsten, iridium, gold, mercury, lead, bismuth, and thorium.

The first observation from this presentation of the detected elements is simply the sheer breadth of them. There is a surprising continuum of elements as well—between no. 22 (titanium) and no. 60 (neodymium) *every single element is present* in the sample, with the exception of no. 43 (technetium)—a highly unstable element prone to radioactive decay. This incredible diversity in materials is quite simply stunning, given the level of purity and purposefulness involved in semiconductor manufacturing— with transistors laid down with nanometer precision. It may suggest that few, if any, of these elements are present as accidents or impurities. More than likely these elements are present for a deliberate purpose, whether as part of building the intricate circuits and logic gates that form the computational machinery of the APU, or for reasons related to the manufacturing process, or something else entirely. By paying attention to the elements the test identified as likely to be present in higher quantities we more than likely avoid elements present accidentally, and also gain a slightly more manageable number of elements to consider. Certain columns of the periodic table are evidently more important to the APUs construction than others—magnesium and barium are light and highly reactive metals both in the same column of the periodic table, both detected in higher quantities. Similar groupings appear at the end of the transition metals (nickel/ palladium, as well as copper/silver/gold and zinc/cadmium), as well as in the post-transition metals (aluminium/gallium/indium in the same column, and tin/lead in their own). Only bismuth, titanium, and chromium stand alone in their own columns, with no paired elements detected in high amounts.

The ICP-MS test itself is highly accurate, able to identify components down to the parts-per-billion scale. It cannot, however, tell us the *quantitative* amounts of each that are present, and it provides no explanation as to why they are present, how they got there, whether they were present in the CPU or the GPU portion of the chip, whether they were present in the outer die casing, present as part of some compound or alloy, or in elementary form. In the real world most metals (with some notable exceptions) are almost never found in pure, elemental form, and great amounts of energy are expended to refine them, removing impurities.

In the sections that follow, we will see some of these processes and the vast amounts of energy involved, building up a picture of the incredible resource intensity of this videogame console. Each of the sections that follow are dedicated to one of these elements present in higher concentrations. In each I look for functional explanations for their presence—why these might be used, or useful to the device—and identify known and suspected environmental harms that sourcing these elements produces, from greenhouse gas emissions to solid waste burdens to toxic tailings and other impacts on the landscape. I used literature searches combining the element name with words like "mining", "extraction" or "environment" to identify these issues. These elements can be read in almost any order, or treated as a reference guide for future work. The periodic table of torture can, I hope, be used as a starting point for similar work on other digital media devices and the results of similar testing procedures.

In this chapter I frequently refer back to Norgate et al.'s (2007) pioneering study of the environmental burdens of key mineral refining processes and the figures they provide, allowing us to identify the environmental harms associated with the energy used in the mining and refining process of many common industrial metals. Norgate et al. (2007) examined the raw energy needed to produce nickel, copper, lead, zinc, aluminium, titanium, steel and stainless steel – metals (or metal alloys in the case of steel and stainless steel) all of which are, with the curious exception of iron, present in higher concentrations in the PS4 sample analyzed. I adopt some common acronyms drawn from Norgate et al. (2007) with their paper offering estimates for both the 'gross energy requirement' (GER) for the production of a ton of a given metal, and its 'global warming potential' (GWP) in terms of kilograms of $CO^2$ equivalent emissions per kilogram of the refined metal—measuring from the point of being dug out of the ground, to leaving the factory gate. The GER figure for each metal remains relatively stable, based on a combination of technological processes and their efficiency, and the theoretical limits of a maximally efficient process, determined by the physical and/or chemical nature of the process itself. In contrast, the GWP figure is a combination of the emissions from generating the energy needed for the processes (often through burning coal or gas, with many smelting reactions only taking place at extremely high temperatures) as well as from fugitive emissions like sulphur dioxide (a major component of acid rain) generated during these transformations.

To practice truly ecological thinking we need to travel where these elements lead us—over the oceans and hills, under mountains, back into and out of the factories and long assembly lines where incredible resource intensity bears down. These are locations all too often hidden from view by concrete walls and chain-link fences as much as by sleek marketing facades, and glossy boxes beckoning us. We must work imaginatively to overcome the barriers to making visible these harms, the pains, the torture inflicted of the world so easily obscured by the tactile seductiveness of brushed aluminium cases, beautifully injection-molded plastics, smooth glass surfaces, and the sheer and immediate joy of switching on our favourite game.

## ELEMENT NO. 12: MAGNESIUM (MG)

The first element on the periodic table that the ICP-MS test detected in substantial quantities was magnesium. What is magnesium doing in the APU of the PlayStation 4, how did it get there, and what sorts of harms may have occurred along the way? It's worth noting some features of magnesium itself—it is the lightest of all the metals that can be used for structural applications, it is quite ductile (bending without breaking), it conducts electricity and heat well, however it also oxidizes violently at high temperatures, and is mostly used in alloys with other metals. A good summary of the applications of magnesium in modern industry is offered by Monteiro (2014: 230) who raises the good thermal conductivity of magnesium that 'makes magnesium alloys a better choice in electronic appliances to dissipate heat generated by electronic circuits.' He also notes that magnesium has an inherent property that allows it 'to shield internal technology from electromagnetic interference and radio frequency interference (EMI/RFI)' (Monteiro, 2014: 234). Monteiro also gives an example of the application of magnesium alloy AZ91D by Phillips Magnesium Injection Molding which produces 15 different components made out of AZ91D including RAM and CMOS chips, screens, chassis' and cases for various parts, and a host of others (Monteiro, 2014: 235). An educated guess might then be that some magnesium is likely present as a structural component of the die of the APU (the hard outer shell that contains the billions of transistors and microscopic circuitry) possibly to assist in dissipating the heat it generates, perhaps assisting with providing insulation from electromagnetic interference. There may be other uses beyond these which I have not been able to identify as well. Regardless of why exactly it is there, it is present and as the test indicates, in substantial quantities.

The US Geological Survey (USGS) estimates that deposits 'from which magnesium compounds can be recovered range from large to virtually unlimited and are globally widespread' (USGS, 2019: 103). Magnesium and magnesium oxide can be recovered from seawater and other salt water sources, with only one primary magnesium producer in the US operating in Salt Lake City producing primary magnesium (USGS, 2019: 100, 102). It can also be mined from the naturally occurring mineral magnesite, with China the leading global producer by this method.

Chinese mining of magnesite typically occurs around the city of Haicheng in the province of Liaoning, which has been described as 'the magnesium capital of the world' (Wang et al., 2015). They explain that as results of the low technological complexity and haphazard mining done in the region, the area faces 'an apparently intractable ecological disaster' (Wang et al., 2015: 90). The mining done in this region (the USGS mentions the use of explosives) results in the release of magnesium oxide dust which 'can react with carbon dioxide and water in air and soil, changing into magnesium carbonate ($MgCO_3$) and magnesium hydroxide ($Mg(OH)_2$), and in turn causing soil pH to increase' (Wang et al., 2015: 90). In other words, the soil becomes more caustic, like baking soda, and in addition to this the dust forms a crust on the surface. According to Wang et al. (2015: 90) this has 'a marked negative effect on the soil hydrological cycle, leading to reduce penetration and to increase rainfall runoff, in turn affecting wider eco-environmental processes.' The scale of the area affected by dust from magnesite mining in Liaoning province is staggering, with one study estimating the area to be around 78,000 hectares. If you can imagine 78,000 football fields covered to varying degrees by a hard crust nearly the pH of baking soda then you begin to get a sense of the possible scale of the issue. Fortunately, hints of improvements appear in the USGS report on magnesium prices from 2018:

> policy changes in China affected prices and availability of all grades of magnesia in the world market... magnesia plant shutdowns in 2017 and 2018, ordered by the Government of China for environmental concerns, resulted in limited supplies and price increases. The Government of China also

restricted the use of explosives and certain equipment in magnesite mines in some areas, resulting in shortages of raw material for some magnesia producers.

The report also mentions the planned consolidation of small-scale magnesite miners in the area, potentially addressing the haphazard mining highlighted by Wang et al. (2015), however these were not yet completed as of 2018 (USGS, 2019) and I have yet to find an update.

Unfortunately magnesium dust is not, however, the only ecological harm connected to this mineral and its role in manufacturing digital devices. Mining of magnesite is typically just the first stage in the process of transforming it into a useable commodity, and the refining process to turn magnesite (or in other processes, salt water) into useable magnesium requires significant amounts of thermal or electrical energy. The Wikipedia description of the 'Pidgeon Process' describes transforming magnesium oxide into magnesium through high temperature combination with silicon, and an average requirement of about 35–40 MWh per ton of metal produced. Depending on how the energy for heating is sourced (e.g. from the burning of coal or gas), this is likely to produce a significant amount of carbon emissions. It is worth noting that the energy for this refining process *could* be made relatively clean. Ross Garnaut (2019) has laid out a plan for Australia to move its heavy industry and metals sector over to renewable power, creating products for export like "green" carbon-neutral steel. It is possible to clean up these processes, they do not have to be (at this stage in the process, at least) inherently polluting and hazardous.

There is also at least another clue as to the harms that might be involved in the use of magnesium in electronic devices, with the use of sulfur hexafluoride ($SF^6$)—a potent greenhouse gas—standard in the casting process. A US Environmental Protection Agency (EPA) report by Bartos (2002) explains that 'the magnesium industry commonly uses sulfur hexafluoride ($SF^6$) to prevent the rapid oxidation and burning that occurs when the molten metal directly contacts air. Many companies apply a cover gas of dilute $SF^6$ mixed with dry air and/or carbon dioxide ($CO^2$) to maintain a protective layer on the molten metal surface.' Monteiro (2014: 236) discusses the development of an "eco-magnesium" by South Korean electronics manufacturer LG, through a casting process eliminating $SF^6$. Monteiro (2014: 236) notes that 'as a result, LG plans a reduction in greenhouse gas emissions by a factor of approximately 24,000 during the die casting process without affecting product quality,' so improvements are possible.

B. J. ABRAHAM

When we think ecologically we often find the bounds of our analysis exceeds our initial focus on just one element or object, encountering other potent greenhouse gases and other environmental burdens in the process. Even though the amounts of magnesium used in any one PS4 is likely quite small (its more substantial application being in the automotive sector) as this brief analysis has shown, its presence in the PS4 indicates that the games industry is not free from entanglements with these global industrial processes and the harms they cause. Much more work remains to be done to figure out the extent of these entanglements, and what can be done about them.

## ELEMENT NO. 13: ALUMINIUM (AL)

The thirteenth atomic element, and the third most abundant element in the earth's crust, (Australian aluminium council, 2020) aluminium has been used since the invention of refining processes for it in the 1880s. A ubiquitous feature of modern manufacturing, billions of aluminium cans are produced for beverages every single year and there are a wide variety of applications in other sectors including, as our sample shows, in digital devices even as it pales by comparison. A lightweight, structural metal that is not all that different from its close relative magnesium (having just one proton difference between them placing them adjacent on the periodic table of elements), aluminium conducts electricity, however not nearly as efficiently as copper, gold, or silver which are generally preferred in the construction of electrical circuits. In their study of the potential for end of life electronics to become a significant source of recycled metals, Oguchi et al. (2011) identified median rates of aluminium in printed circuit boards of video game consoles at around 40 g/kg—or only about 4% by weight— nothing to really get excited about. Even though aluminium is only a tiny part of the PS4's APU—potentially part of a structural alloy with titanium (see element no.22)—it is, however, one of the dirtiest metals to produce and its presence even in small quantities should not go overlooked.

Aluminium is refined in two stages from the highly common mineral bauxite, with world reserves estimated to be between 55 to 75 billion tons (USGS, 2019: 31). Bauxite is largely found in tropical and sub-tropical parts of the world, with China and Australia together producing about half of the world's supply in 2018. The mineral is first excavated from the ground, crushed, and then mixed with caustic soda in a pressure vessel at about 180 °C, which then produces an aluminium oxide (alumina) when it is filtered and cooled. The alumina resulting from the first stage of refinement then must undergo the Hall–Héroult process, involving dissolution of the crystallized material into a molten-salt bath at around 950 °C, which helps to lower its melting point, and allows the pure aluminium to be extracted via electrolysis. The two electrodes that sit in the molten-salt bath acting as the catalyst for the reaction are typically refined carbon, and a byproduct of the reaction is carbon-dioxide. Paraskevas et al. (2016: 209) summarizes the process's environmental impact:

> Primary aluminium production is one of the most energy intensive materials, requiring on average 66MJ of energy per kg in 2012, of which 80% as electricity in the Hall-Héroult process. The Greenhouse Gas emissions associated with the 2007 global aluminium cycle account for 0.45 Gt $CO^2$ eq., or approximately 1% of the global GHG emissions.

Das (2012: 286) also estimates the emissions from aluminium production around the same time, reaching an even higher figure (0.5 Gt $CO^2$ equivalent) and suggesting it accounted for 1.7% of global emissions, only slightly behind Iron and Steel production at 4% of global emissions. These emissions occur despite producing only 2.8% of the volume of iron and steel, making it far more emissions intense by weight. These figures should cause discomfort, and prompt us to consider much more carefully its use in various products (digital or otherwise), at whether it is necessary, and what might be done about it. Encouragingly, this emissions intensity is not set in stone. As Norgate et al. (2007: 845) highlight, the GWP for various metal refining (including aluminium) will fluctuate depending on the source of power for these energy intense processes. In a graph comparing the different energy sources and their GWP, use of hydroelectricity halves the emissions of this industrial process (Norgate et al., 2007: 845). Norgate et al.'s (2007) modelling shows that, out of the 8 metals compared, aluminium had the second highest GER and GWP, second only to titanium, and which is produced in much smaller volume. (See element no. 22 for details) Here too, Garnaut's (2019) analysis of the potential for countries with high-renewable potential (like greatly solar-endowed Australia) to transition primary metals manufacturing entirely to renewable energy is a huge opportunity for further reductions in emissions, become a renewable "superpower" in the process. While technical challenges persist in the implementation, the possibility is tantalizing.

Also of concern is the gradual depletion of the richest, highest grades of bauxite ore, which mean greater volumes of processing for lesser amounts of aluminium. This is a problem facing many mining operations, as Norgate et al. (2007: 844) highlights:

> as higher grade reserves of metallic ores are progressively depleted, mined ore grades will gradually decrease. This reduction in grade will have a dramatic effect on the energy consumption and accompanying greenhouse and acid rain gas emissions from metal production processes.

The use and prioritization of recycled aluminium over 'virgin' metal presents an opportunity, as aluminium has at least one benefit over nonmetallic materials (such as wood and plastics) in that it is 100% recyclable, with no loss of material in the process. There is substantial benefit to the recovery of aluminium from scrap, as once produced it no longer needs to undergo the energy-intense electrolysis of the Hall-Héroult process. A byproduct of this primary production process is a substantial quantity of so-called 'red mud' tailings. For every ton of aluminium produced, approximately 1–1.5 tons of bauxite tailings are generated, taking the form of a highly caustic waste water containing sodium hydroxide. These are usually collected into large ponds requiring careful management, with failures resulting in overflows and other discharges of this strongly alkaline material into the environment with disastrous consequences. Accidents have occurred in both Italy and Hungary, with shocking consequences for both human and non-human life (FranceTVInfo, 2013; Gura, 2010). Recycling and recovery of aluminium from cans and other scrap avoids adding to these other waste burdens, already contributing to a large portion of aluminium manufacturing, with the USGS noting that 'in 2018, aluminum [sic] recovered from purchased scrap in the United States was about 3.7 million tons, of which about 58% came from new (manufacturing) scrap and 42% from old scrap (discarded [aluminium] products)' (USGS, 2019).

In terms of the end-of-life cycle of products like the PS4, the presence of aluminium presents an extremely low environmental risk, being inert, with little or no risk of leeching into groundwater, or causing other contamination. Beverage cans and car bodies made primarily from aluminium are a far cry from the amounts inside the PS4 APU. The entire chip itself weighs at most a few grams, and much more valuable and precious metals exist within it that are worth prioritizing over the aluminium like the platinum group metals, or PGMs (see element no. 46 palladium). When considered in the context of aluminium's overall greenhouse impact, even given how much material may have been used across the roughly 100 million PS4s that have been sold, this will probably rank lower on our priority list than others. Yet it serves to remind us that these devices do not appear without creating substantial burdens for the atmosphere and environment, with much work still to be done on mitigating these emissions. There are still things that may be done to ensure it does not add further emissions and spread further harm. Hardware manufacturers could insist on only green virgin aluminium or 100% from recycled scrap, and immediately guarantee a lower overall emissions profile for devices. This may be a difficult task, given the complicated and opaque nature of modern supply chains, suppliers hard to audit, all in the context of potential purity requirements of semiconductor manufacturing. But if possible, steps like these would be a welcome aid to our chances of averting climate disaster.

## ELEMENT NO. 22: TITANIUM (TI)

Atomic element number 22 is titanium, another well-known metal, with a reputation for high strength applications such as in drill bits, and more extreme applications like the crash-resistant black boxes of airplanes. Its appearance in the PS4 APU is perhaps somewhat surprising. In industrial applications, titanium's main advantage is its high strength to weight ratio. There does not seem to be much need for this in the PS4 APU, however, an alternative explanation for its presence is in the form of the compound lead zirconate-titanate (or PZT) which is 'the most widely used piezoceramic material' (Acosta et al., 2017: 1). PZT can form part of volatile memory systems, such as DRAM, and its presence may be explained by the various working memory locations inside the APU. There is also the possibility that titanium is present in the PS4 APU as part of a common metal alloy made from Titanium, Aluminium and Vanadium known at Ti-6Al-4V (being 90% titanium, 6% aluminium and 4% vanadium) is a commonly used alloy. While aluminium was detected in significant quantities in the sample (see element no.13), vanadium was only detected in trace amounts, much as we might expect if it only formed a mere 4% alloy. Ti-6Al-4 V is the most widely used titanium alloy, making up as much as half of all titanium in use. Titanium, however, has very poor thermal conductivity meaning that it is perhaps not a great material for construction of the die. Heat-dissipating purposes are usually more of a priority in CPU/GPU construction than high-strength, with the mostly rigid silicon PCB that the chip is affixed to providing enough structural support. It seems plausible then that there might be both PZT and structural uses of titanium in the PS4 APU.

Titanium is another metal with very low reactivity, making mining for it less contaminating and of a slightly lower environmental impact than some of the other elements found on this list. For instance, in a study of titanium mine tailings conducted between 1984 and 1987, Olsgard and Hasle (1993) looked at the impact of the depositing left over sediment into the ocean around the Jøssingfjord area in Norway. Over several years of observations about the effects of these tailings, which were mostly silt particles of a range of sizes and that included elevated levels of titanium dioxide ($TiO^2$), they concluded that 'the disposal of mine tailings does have an adverse effect on the benthic fauna' (Olsgard & Hasle, 1993: 190). Their study did suggest, however, that this effect was less to do with chemical reactions with $TiO^2$ itself, but rather the effect of the small particles of silt that formed the tailings, becoming deposited on the sea floor, smothering the living inhabitants of the seabed. A complicated picture emerges from the study of different species adaptability to being 'buried' under different types of silt, once again emphasizing that even lesser impact does not mean no impact whatsoever:

> Most studies of [titanium] mine tailings describe the waste as almost or entirely non-toxic to marine biota. This is the case here as there are almost no toxic discharges. Thus the main adverse biological effects are likely to be connected to the rate of sedimentation of tailings, and not to the presence of tailings in the sediments itself. Seabed organisms are smothered and destroyed wherever there is a rapid, heavy settling of tailings, with less impact in areas less affected by sedimentation. (Olsgard & Hasle, 1993: 206)

This reinforces the importance of taking an ecological perspective rather than a singular focus on, for instance, the toxicity of various elements and processes. Examined in isolation, titanium mining appears non-toxic and to have a lesser impact, but if the by-products of titanium are disposed of in ways that smother aquatic life, or that require transportation powered by fossil fuels—the non-toxicity of the by-products of titanium mining does not give a free pass.

There is also the associated greenhouse gasses emitted by titanium mining, and Norgate's (2007) study gives titanium the highest, that is *worst*, figures for both the GER and GWP of any metal in their study. Refining processes for titanium require almost 350 MJ of energy expended to produce 1 kg of titanium, and emitting as much as 35kgs of $CO^2$ equivalent emissions (Norgate et al., 2007: 844). Titanium's most important property—its great strength—is also why it is much harder to work with than other metals, requiring higher temperatures and more energy, whether refined through the Kroll process (around 1000 °C) or the Becher process (temperatures of 1200 °C). How this heat is generated—by burning fuels or through electricity will greatly determine the overall emissions intensity. Both processes also involve the use of other chemicals either for the removal of non-titanium elements from the ore, such as iron (done via chlorine gas or sulphuric acid), which require handling to avoid environmental releases.

Even though the ICP-MS test indicated it is likely to be present in substantial quantities, given the sensitivity of the test the amounts we are dealing with in the PS4 APU are in the order of grams, perhaps less. If it is used in memory applications (such as DRAM) then it will almost certainly have been of a high grade of purity, requiring yet more energy to purify. Even these small amounts can add up to significant emissions embodied in each console with the scale of modern consumer electronics manufacturing and present real environmental harms which deserve both acknowledging and addressing.

## ELEMENT NO. 24: CHROMIUM (CR)

Chromium, the 24th element of the periodic table, is a shiny metal that can be polished to an almost mirror-like reflection. Chromium is used widely for its resistance to abrasion, oxidation (i.e. rust), and for waterproofing leather. Chromium's many useful properties naturally lend it utility for manufacturing digital devices as well, and it is most crucially used in the process of photolithography, enables modern CPUs and GPUs manufacturing. The process, invented almost simultaneously in the late 1950s by two electrical engineers—Jack Kilby at Texas Instruments, and Robert Noyce at Fairchild Semiconductor—was an extension of the existing process of manufacturing transistors. It remains conceptually the same today, despite involving greater automation and operating at ever shrinking scales. The process involves the layering of different semiconductor materials like silicon, often doped with various other elements like boron (element no. 5—not considered here, but also present in the sample) that lend it the necessary properties that create the electrical switch that we call a transistor. Over the top of these, layers are arranged in various patterns using different types of masking materials, which resist being eaten away during the etching process. Etching involves chemical baths that eat away exposed elements, leaving only the masked patterns creating circuits. One of these masking layers may well be a chromium layer, with its useful corrosion resistance.

The presence of Chromium in the PS4's APU may be partially explained as resulting from this photolithography process, as the microscopic protective chrome layer may remain following the etching process. Another possibility may be the presence of stainless steel in the APU, perhaps as part of the chip's outer casing. Chromium is the main non-steel component of stainless steel (alongside nickel—see element no.28, also present), giving the highly useful alloy its resistance to scratching, rusting and its polished look—but this utility here seems questionable.

Chromium is typically refined from the ore chromite, in which chromium exists in a compound as chromium oxide ($Cr^2O^3$)—also referred to as chromium (III)—and which is 'among the ten most abundant compounds in the Earth's crust' (Guertin et al., 2004: 4). Guertin et al. in (2004:4) provides the following helpfully summary:

> Chromium is rarely found as a free metal in nature. A clean surface of Cr metal reacts strongly with the atmospheric oxygen. However, the reaction stops quickly owing to the formation of a strong, dense, and nonporous Cr(III) oxide surface layer. (Guertin et al., 2004: 4)

It is the atomic structure of chromium and its chemical properties when combined with other elements that give it such a useful variety of possible 'oxidation states'. Chromium (III) (or trivalent chromium) is abundant in nature, and has three fewer electrons than the more commonly used industrial form, which is chromium (VI) (or hexavalent chromium). Hexavalent chromium is a known carcinogen, and strictly regulated in most of the world, presenting a major human health and environmental hazard and often results as a waste by-product (EPA 2018). In a study of the effects of chromium (III) and chromium (VI) in plants, Shanker et al. (2005) explains that: 'Cr(VI) is considered the most toxic form of Cr.' Shanker et al. (2005) notes that the 'presence of Cr in the external environment leads to changes in the growth and development pattern of the plant' (Shanker et al., 2005: 742). These changes are invariably detrimental, including 'alterations in the germination process as well as in the growth of roots, stems and leaves, which may affect total dry matter production and yield' (Shanker et al., 2005: 739).

The EPA (1998) toxicology review of chromium lists dozens of studies involving both humans and animals exposed to chromium, and found significant increases in lung cancer in workers exposed to both trivalent and hexavalent chromium, increases in health problems in villagers exposed to it via groundwater, and numerous other health issues from exposure. More detailed discussions of the environmental impact of chromium on the environment can be found in Losi et al. (1994), or the EPA's Chemical Assessment Summary for chromium (EPA, 1998), as well as the Independent Environmental Technical Evaluation Group's *Handbook on Hexavalent Chromium* (Guertin et al., 2004). Suffice to say, however, there are serious environmental and public health challenges presented by the use of this substance.

Perhaps because of the seriousness of the environmental and human health challenges presented by the hexavalent form of chromium, the overall greenhouse gas emissions of chromium mining and refining tend to take a backseat in the literature. The chrome plating process in particular receives the lion's share of attention, as it involves a form of electrolysis and requires significant amounts of electricity. However, as the USGS notes (2019), there are no identified substitutes for chromium in either stainless steel production, chrome plating, and other durability enhancing applications, most of these uses will not be significant to manufacturing digital technology however. In conclusion, we are left with some ambiguity about the overall contribution to emissions from the demand for chromium, from semiconductor manufacturing and devices like the PS4 with few replacements or alternatives. The same cannot be said, however, for the toxicity of chromium in its hexavalent form, the other environmental impacts of which are now well known.

## ELEMENT NO. 28: NICKEL (NI)

Nickel is another highly abundant and commonly used metal, and according to the US Geological Survey 'about 65% of the nickel consumed in the Western World is used to make austenitic stainless steel' (USGS, 2021a).

In digital applications, nickel is critical to a number of different recharge-able battery chemistries, from the nickel-metal hydride (NiMH) battery which offer a rechargeable alternative to single-use 1.5 V alkaline batteries, as well as being a part of some lithium-ion (Li-ion) batteries, with cathode material frequently an alloy of lithium, nickel, cobalt and aluminium oxide. There is no room or need for batteries here, however, as power is supplied to the APU from the PS4 power supply.

Nickel has as a range of other applications in non-ferrous alloys with aerospace and vehicular applications, as well as being useful in certain high temperature environments. As hot as the PS4 APU gets, it never approaches the sort of temperatures these so-called super-alloys are designed for, such as turbine blades withstanding up to 1000 °C or more. There is also scant mention in the literature of other uses that might be involved with digital media devices beyond nickel's use in batteries. My primary assumption, then, was that nickel is present in substantial quantities in the PS4 sample because it is part of a non-ferrous alloy, either in the die's casing or some other structural part of the APU. It is certainly not due to the use of stainless steel, however, as despite the test detecting iron (element no. 26) it was *not* detected in significant quantities. As iron is the main component in stainless steel—at the very minimum 60%—it should have turned up in substantial quantities as well. The only other uses for nickel I have found in the literature are so-called 'thick film conductors' (Kasap & Capper, 2017: 711) for certain niche semiconductor applications.

Nickel is typically refined from two main types of ore (sulfide and laterite) which each require different processes to extract through various involved and energy intensive processes. For the simpler process, sulfide ores are extracted pyrometallurgically (via blast furnace) after a series of process of flotation and concentration, and this has been how most nickel has historically been refined. However in recent years, an economical process enabling the extraction of nickel from laterite ores has arisen, and represents a growing portion of ore mined. This process is much more complicated, involves a number of configurations, but generally makes use of a hydrometallurgical process of high-pressure acid leaching (HPAL). Both processes take a significant amount of energy, and according to Norgate et al. (2007) nickel has the third highest GER and GWP of the most commonly produced metals, varying slightly depending on the method. For blast furnace refining, Norgate et al. (2007) estimated 114 MJ of energy for 1 kg of nickel produced, while for pressure-acid leaching, they offer the substantially higher figure of 194 MJ of energy. For a sense of just how much energy we are talking about here, we can convert megajoules to kilowatt hours: 194 MJ of energy, if perfectly converted to kilowatt hours of electricity, is a little under 54 kWh, coincidentally the same capacity as the standard range Tesla model 3 in 2020. With that much stored energy, the car is able to travel about 400 km–the same amount of energy is used to produce a single kilogram of nickel as it takes to travel 400 km in a state-of-the-art electric vehicle.

Beyond immediate energy intensity, nickel is mined in several locations around the world with impacts on their surroundings. The biggest and most well developed fields are in Canada, however nickel mines also exist in Russia, New Caledonia, Australia, Indonesia, Cuba, Colombia and China (Mudd, 2010). Predictably, there are substantial historical and contemporary environmental problems that have accompanied nickel mining: sulphur dioxide ($SO^2$) releases into the atmosphere have contributed substantially to acid rain, particulate matter reduces air quality, and a variety of heavy metals released also contribute to environmental degradation surrounding nickel mining, refining and smelting operations. Mudd (2010) offers short summaries of the history of each of the main fields listed above, and includes some relevant historical context. The largest deposits, the Canadian Sudbury nickel fields, over their history have created significant unwanted impacts:

> The early practice of heap roasting prior to smelting led to severe local $SO^2$ emissions and impacts, a significant issue noted by the Royal Ontario Nickel Commission in 1917...Despite the closure of heap roasting...in 1928 and continued evolution in smelting and refining practices over time, the increasing scale of Sudbury's production led to substantial cumulative environmental impacts by the 1960s over a wide area. The impacts included heavy metal soil contamination, acid rain linked to $SO^2$ emissions (~1.5 to 2.7 Mt $SO^2$/ year), acidified wetlands, biodiversity declines (especially fish), vegetation dieback and heavy soil erosion. (Mudd, 2010: 15)

With the rise of the first wave of environmentalism in the 60s and 70s, both public and governmental awareness of these sorts of environmental harms arising from mining led to improvements. In Ontario (where the Sudbury field is located) regulators specified maximum emissions levels for $SO^2$ which 'effectively forced production to match emissions levels until smelting and $SO^2$ capture technology and infrastructure caught up...

Emissions limits and ground level concentrations are still being reduced over time. Over 2001 to 2005, Inco's $SO^2$ emissions from Sudbury declined from 230 to 194 kt $SO^2$/year' (Mudd, 2010: 15).

In a highly encouraging turnaround that should spur us to redoubling of regulatory efforts, Mudd (2010: 15) informs us these changes led to 'the gradual recovery of the surrounding Sudbury environment. Acid rain problems and the numbers of damaged lakes have significantly reduced and fish stocks have increased substantially.' Less encouraging are reports of the continuing degradation that persists in regions surrounding yet-unremediated and still poorly regulated Soviet-era nickel mining operations in the Russian Taimyr and Kola peninsulas: 'Both fields are [still] considered to be major sources of mercury contamination in the Arctic. The Kola Peninsula in particular is renowned for contributing to pollution problems across northern Europe' (Mudd, 2010: 19). The difference in outcomes from areas with strong environmental regulations and compared to those without is reflected in the marked improvements on previously declining environmental trends which might guide our responses to these types of issues in the future, with their persistence in certain regions serving as a warning to the continued dangers of unregulated, unfettered extractivism. This is not, however, to fully excuse ongoing (lesser) emissions, but merely to point to achievable concrete reforms.

One last encouraging detail is the high degree of recyclability of nickel products, with the Nickel Institute (2021) stating that:

> Most nickel-containing materials are fully recyclable at the end of the product's useful life; indeed their high value encourages recycling. This, in turn, lessens the environmental impact of nickel-containing stainless steels by reducing both the need for virgin materials and the energy that their production uses. For example, the amount of stainless steel scrap currently being used reduces the energy required for stainless steel manufacture by around one-third over using 100% virgin materials.

Similar to aluminium and titanium then, virgin nickel remains both literally and figuratively costly to produce, with a high emission burden. Greater use of renewables in the energy mix for these refining processes, and a greater use of recycled materials would see significant environmental improvements. Whether recycling the nickel in smaller electronics devices like the PS4 APU could become economical remains an open question, one which we return to in the discussion of the platinum group metals (see element no. 46).

## ELEMENT NO. 29: COPPER (CU)

Chances are good that if the average person on the street was asked "what metal is most important for computers?" their answer might be copper. From the earliest powerlines to the nanometer circuits that make modern CPUs as powerful as they are, copper is a ubiquitous and indispensable material. Useful for making electrical circuits, wires, motors, communications networks, and a host of other digital technologies, copper has been critical in enabling the life and habits we have come to associate with the twenty-first century. The USGS explains the importance of copper:

> Because of its properties, singularly or in combination, of high ductility, malleability, and thermal and electrical conductivity, and its resistance to corrosion, copper has become a major industrial metal, ranking third after iron and aluminium in terms of quantities consumed. (USGS, 2021b)

It also notes that 'electrical uses of copper' which includes the tightly wound wire inside electrical motors and generators, as well the powerlines that electricity is transmitted through and into our homes, all together 'account for about three quarters of total copper use' (USGS, 2021b). Copper is essentially the number one power metal. Copper also represents the connecting strands that knit together the countless capacitors, transistors, diodes, MOSFETs, CPUs, RAM and other components that gets assembled, packaged, shipped, placed on a shelf, and sold all around the world as a PS4 or similar game console. Its use in the ultra-miniaturized circuity inside the sealed APU is what surely accounts for the majority of its inclusion in the sample.

As a metal element that can be found in usable form right in the ground, copper has a long history of use, its discovery enabling the bronze age from around 8000 BC. The amount of copper mined in 2018 was over 20.6 million tons, concentrated in a handful of counties: Chile, Peru, China, The United States, and The Democratic Republic of Congo each producing over 1 million tons. Chile in particular accounted for almost a third of the world's total, bringing 5.8 million tonnes of copper to the surface, and with an estimated 170 billion tonnes in reserves still yet to be dug up, about 40% of the world's total deposits (AmCham Chile, n.d.). If you recall the spectacular rescue in 2010 of 33 Chilean miners who were trapped underground for almost 70 days, these miners were working in one of Chile's many copper mines, and it remains highly dangerous work.

In Merchant's (2017) *The One Device*, the author travels to Chile to see inside one and is staggered by the conditions, and the risk faced by miners on a daily basis, the frequency of injury and death for those who go down the mines. A "Where are they now?" style story about the Chilean miners five years on from the ordeal found many of them struggling with lingering trauma from being trapped underground for so long, even as the political and media spotlight moved on, many of those same miners finding the decision in 2013 to not prosecute the mine's operators bitterly disappointing (Pukas, 2016). Copper and other mining makes up a huge part of Chile's economy, and almost half of its exports, and still has a mixed track record at safeguarding mine workers.

Like other metals, copper is highly recyclable, with no losses in the process. What complicates this otherwise rosy picture, however, is the difficulty in isolating copper from other valuable metals in electronic devices. A 2016 report by Greenpeace identifying the resource intensity in the communications technology sector outlined a number of challenges facing recycling, particularly for smaller devices such as smartphones and tablets, in large part associated with costs of collection, transport and logistics relative to the trace amounts recovered from each device. Because of this, estimates for the amount of these devices that actually reach suitable recycling collectors are at best less than 50% and 'probably even below 20%' (Manhart et al., 2016: 41). Despite being the most energy efficient devices at runtime, as we saw in prior chapters, phones are among the hardest to recycle.

The same report provides further details on the recycling process. Once a device has been collected there are significant labour costs involved in getting the device ready to be processed: an initial depollution step involves removal of batteries, external cables, etc., with some recyclers also using this stage 'to screen incoming WEEE for devices suitable for repair and reuse' (Manhart et al., 2016: 43). In countries where labour is expensive, such as in the EU, these steps are often done in ways that do not maximize recycling potential. The actual end processing of e-waste is done in one of a couple of ways, 'by mechanical processes (e.g. shredding), manual operations (dismantling) or a combination of the two' (Manhart et al., 2016: 44). The advantage of shredding is that it can be automated and involves very little human labour, however, this method has 'the disadvantage that they lead to losses of valuable metals such as gold, silver and palladium' (Manhart et al., 2016: 44) (See element no.'s 46, 47 & 79) This is a bit

concerning because while the world has close to 830 billion tonnes of copper in reserves (USGS, 2019: 52), the estimated reserves of platinum
group metals (e.g. platinum, palladium, iridium, etc.) are far less: 'World
resources of PGMs are estimated to total more than 100 million kilograms' (USGS, 2019: 24). The platinum group metals (which I touch on
in element no. 46—palladium) are orders of magnitude rarer than copper
and much more important to conserve. They are also substantially more
valuable than copper, and thus recyclers have sizeable economic incentives
to capture them from this waste stream. The report notes the challenge
that recycling small devices presents, suggesting 'there is no perfect recycling path for smartphones and tablets' (Manhart et al., 2016: 45). They
also note that recycling other more common metals like steel and aluminium (as I suggested earlier for its importance in reducing GWP burdens) 'are only relevant for devices that undergo mechanical pre-treatment.
[…] On the other hand the practice of feeding handsets directly into secondary Cu-smelters leads to losses of aluminium and iron as these metals
move into the slag phase of these smelters' (Manhart et al., 2016: 45).
The report also raises the practice of exporting these waste streams to
countries (most notably Ghana) where 'various process-steps by these
recyclers (e.g. the open burning of cables to recover copper) are associated
with extreme levels of pollution' (Manhart et al., 2016: 45).

Given these competing factors it is difficult at present to say which metals to prioritize in recycling flows, potentially varying from device to
device. Copper production has a relatively low GER: 33 MJ/kg of copper
for smelting/electrorefining, or 64 MJ/kg for heap leaching processes
(leeching copper from ore via sulfuric acid). Because of these (comparatively) low GER figures, the GWP is also lower than other metals, being
3.3 kg and 6.2 kg $CO^2$ equivalent for the two method respectively—orders
of magnitude lower than those discussed previously. Still, recycling copper
remains far more environmentally friendly than primary production.
Recycled copper from scrap, 'contributed about 35% of the U.S. copper
supply' according to the USGS (2019: 52).

The risks posed by copper in the natural environment are quite low, as
mentioned above it is readily found in nature in usable forms. The US
EPA has a 'lead and copper drinking water rule' established due to the
widespread use of copper and lead pipes in plumbing uses, with guidelines
stating an 'action level' of 1.3 mg/L of copper in water, much higher than
lead's trigger point at just 0.015 mg/L (EPA, 2008). The much lower

toxicity of copper makes it relatively safe, being an essential element for many organisms' normal function—lobsters and crustaceans for instance use copper instead of iron to transport oxygen in their blood, and the EPA's copper warnings are largely concerned with consequences for aquatic rather than human life, noting that copper 'is toxic to aquatic organisms at higher concentrations' (EPA, 2020). Overall, the picture of copper's environmental impacts comes largely from the sheer scale of production around the world, with a much lower GER and GWP than previous metals described. It is therefore of lesser concern than many of the others in this chapter.

## ELEMENT NO. 30: ZINC (ZI)

In Australia, summer days spent at the beach often involve parents smearing a colourful paste on the noses and cheeks of their children to block harmful UV rays with a thin layer of reflective particles. These particles are a zinc oxide, which creates a physical barrier on the surface of the skin preventing sunburn. For similar reasons of durability to the elements and resistance to rust, zinc is often used in roofing and exterior cladding of buildings. In electronic devices, zinc may be present as part of a Gallium-Indium–Zinc–Oxide (GIZO) compound, a kind of "thin-film transistor" (TFT), most commonly used in LCD and OLED screens as it can be deposited onto layers of glass (Kwon et al., 2008). Other uses perhaps more germane to its use in the PS4 APU are other zinc-bearing compounds—in particular diethylzinc, a highly flammable material used as part of zinc-doping in transistors, and there are some references to Indium-Zinc-Oxide as another potential semiconductor material which could equally explain its presence in the APU (Kasap & Capper, 2017: 1117). By this point in this list we are mostly beyond the types of materials that are likely to be *structural* elements, into the territory of rarer and more exotic elements, and more likely to be part of the computational machinery itself and the complicated manufacturing processes. Zinc and many of the elements that follow, will probably be performing a more specialized role than we have seen previously.

It is worth emphasizing again that the quantities involved in any one PS4 are likely quite small—yet the volume of consoles and other devices being made, and the inevitable waste from the manufacturing process, means even small amounts can add up to substantial demand. Despite being an essential element for the normal functioning of many lifeforms—it's added, for instance to the feedstock of livestock like pigs and often ends up concentrating in environmentally deleterious levels in animal waste (Jondreville et al., 2003)—for uses such as these zinc also has to be mined. Similarly, the nature of modern transistor manufacturing, which operates almost down at the level of single atoms, likely requires a higher grade of purity in its source materials (like zinc) than when used for industrial cladding purposes or zinc as a UV sunscreen. The extra steps required to produce ultra-pure materials will inevitably increase energy intensity, with one manufacturer of "zinc sponge", a raw form of granulated zinc, selling a range of purities, from the basic 99% pure, to the 99.999% pure (American Elements, 2020).

To obtain zinc for manufacturing, it must first be mined from sulphide ores which usually contain an amalgam of lead and zinc, and which can be either refined following concentration of just the zinc portion, the lead portion, or as a mixture containing both. This makes accounting for the GER and GWP for zinc slightly more complicated, helpfully however Norgate et al. (2007) include figures for both. When a concentrated zinc ore is refined by the electrolytic zinc process Norgate et al. (2007) estimate a GER of 48 MJ/kg of zinc produced, and a GWP of 4.6 kg of $CO^2$ equivalent emissions. When refined via the imperial smelting process which produces both lead and zinc at the same time, it results in a slightly lower figure of 36 MJ/kg (GER), and 3.3 kg of $CO^2$ equivalent emissions (GWP). A further additional environmental burden is often created, as the closely related, heavy metal cadmium (see element no. 48) is frequently a byproduct of the refining process. Zinc and cadmium are in the same column of the periodic table, however while zinc is an essential element for life, cadmium is highly toxic. The EPA (2005) toxicology report into zinc notes that it 'is an essential element, necessary for the function of more than 300 enzymes' and that there is generally incomplete data and studies to drawn on for certain human toxicity and carcinogenic levels, however recommendations for maximum exposure levels (in the mg per kg order of magnitude) seem unlikely to affect many except those working very closely with it. Suffice to say, it is not a greatly toxic material, and its environmental impact seems mostly localized at the mining & refining stage of the process, through carbon emissions and lead and cadmium byproducts generated.

## ELEMENT NO. 31: GALLIUM (GA)

The first of what might be considered the 'rarer' minerals, and the immediate next door neighbour to zinc on the periodic table is gallium, with just one more electron, we can begin to imagine why zinc and gallium might work well together in transistor construction given electricity is the flow of electrons. Gallium's presence in the PS4 APU is almost certainly due to its use as a common semiconductor material, having been first used in the construction of transistors in the 1960s with gallium arsenide (GaAs). According to Schulz et al. (2017: H.2) these are still in use today 'in feature-rich, application-intensive, third- and fourth-generation smartphones and in data-centric networks.' An alternative compound, Gallium Nitride (GaN) is used in high powered devices 'because GaN power transistors operate at higher voltages and with a higher power density than GaAs devices' however the voltages applicable are substantially higher than the typical for computing hardware: up to 48 volts in some cases, and 10 amps or greater, too much power for the APU to handle (Schulz et al., 2017: H1). Almost certainly gallium is present in the sample solely as a semiconductor material, part of the transistors inside the APU.

As far as mining for gallium is concerned, it is a relatively rare mineral, found 'most commonly in association with deposits of aluminium and zinc' (Schulz et al., 2017: H.6) and is mostly produced 'as a byproduct of the processing of bauxite ore for aluminium, with lesser amounts produced from residues resulting from the processing of sphalerite ore for zinc' (Schulz et al., 2017: H6). The United States total consumption of Gallium in 2018 was approximately 630 metric tonnes of gallium wafer, and 33 metric tonnes of the pure metal form, of which 100% was supplied by imports (USGS, 2019: 62). The USGS also reported some recycling of gallium, with some gallium wastes 'generated in the manufacture of GaAs-based devices... reprocessed to recover high-purity gallium at one facility in Utah' (USGS, 2019: 62). Low grade gallium in 2018 cost about $200 a kg, indicating that some economic incentive exists for its recycling, however similar difficulty to that described in the discussion of copper exists, as this is still substantially less valuable than the precious metals of the platinum group which routinely sell for tens and even hundreds of thousands of dollars per kilogram. (See no.46 palladium) I have been unable to find estimates for either a GER or GWP values for Gallium mining, but given the low volumes being consumed this is likely to be of a lesser priority than addressing its health and environmental effects.

Gallium has been found to have some health impacts, especially if improperly handled or released into the environment (New Jersey Dept of Health, 2001). On its own, pure gallium is a liquid from 26 °C on up to extremely high temperatures, and can cause corrosive burns according to the New Jersey Dept of Health. Their fact sheet claims that it 'may damage the liver and kidneys... [and] affect the nervous system and lungs' (New Jersey Dept of Health, 2001: 1). Worryingly they also note that 'high exposure to gallium may affect the bone marrow's ability to make blood cells causing anemia' (New Jersey Dept of Health, 2001: 1). In the more useful gallium-arsenide form, Chen (2006: 289) notes that 'GaAs... can be toxic in animals and humans... acute and chronic toxicity to the lung, reproductive organs, and kidney have been associated with exposure.' Chen (2006) also notes several studies measuring longer term exposure in laboratory tests have returned similar worrying results, and discusses the levels of arsenic, gallium and also indium (see element no.49) in waterways at sites in and around the industrial regions of the city of Hsinchu, in north-western Taiwan. Chen (2006) concludes that 'the groundwater levels of Ga, In, and As... indicate that the semiconductor industry in the region has affected groundwater' and notes that 'Hsinchu residents may be at increased risk for immune system diseases and for reduced bloody leukocyte counts' (Chen, 2006: 294).

## ELEMENT NO. 46: PALLADIUM (PD)

A bit of a jump from the previous element, we are now on the next row of the periodic table, by-passing a number of elements detected in the PS4 sample but not in high concentrations. These elements were: germanium (no.32), arsenic (no.33—reinforcing the previous suggestion about the presence of Gallium-Arsenide semiconductors), selenium (no. 34), bromine (no. 35), krypton (no. 36), rubidium (no. 37), strontium (no. 38), yttrium (no. 39), zirconium (no. 40), niobium (no. 41), molybdenum (no. 42), ruthenium (no. 44), and rhodium (no. 45). The current element, palladium is right in the middle of the so-called platinum group of metals, named after its most famous element platinum (no. 78), with platinum and palladium sharing a column in the periodic table.

The platinum group metals (or PGMs) are extremely rare as a proportion of the earth's crust and as a result are extremely expensive, owing to their difficulty to obtain. According to the USGS, in the year 2018 'one company in Montana produced about 18,100 kilograms of PGMs with an estimated value of about $570 million' (USGS, 2019: 124) One year of mining operations for only eighteen tons of material is quite a striking statistic. Part of the reason PGMs are so expensive is that they are extremely *useful*, serving as essential catalysts for a host of important reactions. Since the 1970s, the introduction of ever more stringent emissions laws pushed car makers to include catalytic converters in car exhausts, substantially reducing the smog burden in major cities as a result. The reactions that PGMs like palladium catalyze turns carbon monoxide emissions into less harmful compounds. According to the USGS, 'in 2016, the automobile industry continued to be the major consumer of PGMs. Catalytic converters...accounted for approximately 84% of global rhodium consumption, 67% of palladium consumption, and 46% of platinum consumption' (USGS, 2016: 57.1) The introduction of PGMs into catalytic converters also kicked off a suite of research into the potential and emerging health impacts of these types of metals—a special issue of the journal *Environmental Health Perspectives* from 1975 contains a series of studies on the as-then unknown potential effects of introducing these metals into the environment—however a search of the literature seems to suggest many of the initial concerns have not been borne out, and research into PGMs as a public health issue seems to have petered out not long after.

The uses of PGMs in electronics tend to involve much smaller quantities than in catalytic converters, and due to not being involved in catalyzing chemical reactions are unlikely to escape into the environment as long as proper recycling is undertaken. The economic incentives to reclaim PGMs are among the highest of any metals. The charmingly titled USGS report *Platinum Group Metals: So Many Useful Properties* explains that, 'in the electronics industry, [PGM] components increase storage capacities in computer hard disk drives and are ubiquitous in electronic devices, multilayer ceramic capacitors, and hybridized integrated circuits. The glass manufacturing industry uses [PGMs] to produce fiberglass and liquid-crystal and flat-panel displays' (USGS, 2014: 1).

The most likely explanation for the presence of palladium in the PS4 APU seems to be as part of a hybridized integrated circuit utilizing a kind of 'thick film' technology. The Springer Handbook of Electronic and Photonic Material describes it as an alternative approach to silicon based printed circuit boards, with instead 'the films...deposited by screen printing (stenciling)...with the advent of surface-mounted electronic devices in the 1980s, thick film technology again became popular because it allowed the fabrication of circuits without through-hole components.' (Kasap and

Capper: 707) For much of the early history of electronics manufacturing, most components were mounted with their connecting terminals poking 'through' to the rear of the circuit board where they were then soldered into place—hence the term 'through-hole' component—however the increasing miniaturization of components made this less and less feasible as parts shrunk. The 'thick film' part of this process means a layer is added to either provide resistive or conductive elements as needed between different layers of the board (this layer is not what we might ordinarily consider "thick", only thick relative to the sizes involved). The handbook describes silver/palladium alloys as 'the most common type of thick film conductor... most widely used in the hybrid circuit industry. It can be used for interconnecting tracks, attachment pads and resistor terminations.' (Kasap and Capper: 711) There are also conductor elements made form pure forms of one of the PGMs (like palladium) however the handbook notes that:

> not too surprisingly, perhaps, these are the most expensive of all the commercial thick film conductor materials. Platinum films have a very high resistance to solder leaching and exhibit similar electrical properties to those of the bulk material: a linear, well-defined temperature coefficient of resistance. Platinum films are therefore used in specialist applications such as heaters, temperature sensors and screen-printed chemical sensors. (Kasap and Capper: 711)

As a delicate electrical device with stringent voltage and temperature requirements, the PS4 APU will almost certainly feature a built in temperature sensor to monitor its thermal state. The PS4 console's ventilation holes are quite prone to clogging with dust, and when it is unable to circulate sufficient volumes of cool air over its parts it automatically shuts itself off to prevent damage. Many long term users will have seen the blue system screen alert that appears whenever the PS4 overheats and begins shutting itself down.

According to the USGS, 'consumption of palladium by the electronics industry was 38,600 kg in 2016' (USGS, 2016: 57.2) and it is worth asking how (and where) palladium and other PGMs with 'so many useful properties' get dug out of the ground given their rarity. What environmental costs are we being paid for as part of the eye-watering price (around $76,000 USD per kg in March 2021)? It may come as no surprise that these very rare metals are not found in great concentration anywhere in the world. Because of their density these metals have largely descended into the inaccessible core of the planet in the eons since the earth's formation. The remainder that exists near the surface is scattered like geological-scale dust. Some of the best and highest grades of PGMs like palladium, platinum, rhodium and iridium are in deposits in Russia, South Africa and Canada, with the Roby Zone deposit in Ontario typical of these deposits, containing just a skerrick on a weight-for-weight basis. Lavigne and Michaud (2001) describe ore grades at the Roby Zone as containing just 1.55 grams of palladium per *tonne* of rock, and just 0.17 grams of platinum. A video from the popular science *Periodic Videos* YouTube channel—which features accessible discussions of each atomic element led by Sir Martyn Poliakoff, Professor of Chemistry at the University of Nottingham—cites a range of 350–1400 tonnes of rock removed per kg for many of the Platinum Group Metals (Periodic Videos, 2013). This volume of rock removed during mining is referred to as the solid waste burden (SWB) and is an active consideration in all mining operations, but especially so here. I have omitted discussion of this aspect of mining for the sake of brevity elsewhere, but with PGMs it becomes paramount due to its sheer volume relative to the target metals. Norgate et al. (2007) tallies the SWB for most common metals however the low grades of PGM ore worldwide means that mining for PGMs represent a vastly higher SWB and eventually emissions when considering their transport and disposal. This is typically done by diesel powered machinery—excavators, trucks, and so on. Life cycle assessments of other mining processes, even ones that involve energy intense leaching of minerals, often find that the carbon emissions from the transport of vast amounts of heavy rock and dirt often equal or even eclipse the energy used in other parts of the mining and refining process (see, for example gold, copper and uranium mining in Australia discussed by Haque and Norgate (2014) and Kittipongvises et al.'s (2016) details of the SWB from granite mining in Thailand).

Glaister and Mudd (2010: 446) note that, while it is rare for mines to report SWB figures, one that does 'is Anglo Platinum, reporting in 2008 that their cumulative mine wastes were 730.8 Mt of tailings and 665.4 Mt of waste rock.' Glaister and Mudd's (2010) study provides a wealth of other details about PGM mining, collating and analyzing the sustainability reporting done by PGM miners in South Africa and finding a range of figures for the energy required for both primary (ore production) and secondary (transport) activities. Most figures are somewhere between 100–200 GJ/kg (note the order of magnitude higher than estimates for other metals, all in MJ), however they discuss the eye-watering difference in energy intensity associated with SWB for open-cut and subterranean mines, with the lowest an open cut mine at Mokopane (21 MJ/t rock removed), compared to 'the deep Northam underground mine (~2 km) [which] has the highest mining energy consumption (1244 MJ/t rock)' (Glaister & Mudd, 2010: 444). Their study also shows an upward trend in energy use, and offers a worrying projection for the next 100 years, with two different models coming up with separate, equally imposing figures: 'peak emissions are 49.3 Mt $CO^2$ e/year and 96.6 Mt $CO^2$ e/year, respectively' (Glaister & Mudd, 2010: 446). Whichever eventuates, these are staggering amounts of emissions at a time when we should be doing everything we can to rapidly reduce them.

In addition to the emissions burden of removing so much rock, the lack of environmental controls in some mining areas—particularly the Norilsk region in Russia—results in environmental contamination from PGM mining. According to Greenpeace 'although the area is widely closed for external evaluations, it is ranked amongst the world's 10 worst polluted places by the US-American Blacksmith Institute and the Swiss Green Cross' (Manhart et al., 2016: 18). The same report also notes that 'social tension, labour disputes and unrest have been troubling characteristics of the South African PGM mining industry for several years. The situation peaked on 16th of August 2012 when police opened fire on wildcat strikers in the Marikana area. 34 workers were killed in the incident' (Manhart et al., 2016: 18). Sites of resource extraction have always been contested by workers seeking an equitable share in the fruits of their labour, and highly lucrative PGM mining industry is no exception.

The one upside to the high price of PGMs from an environmental perspective is that recycling becomes highly attractive, and these metals are often prioritized over cheaper ones, despite being harder to recover and the tiny amounts involved. Unlike some of the more common elements described here, global deposits of PGMs are easily depletable, especially if demand continues to rise. Thankfully, the transition from internal-combustion engine vehicles is likely to reduce demand for PGMs for catalytic converters. Likewise, better recycling of e-waste and the capture of even small amounts of PGMs is going to be crucial if the world wants to keep on using and producing electronics like the PS4, even with its relatively small amounts of palladium and other PGMs. The USGS minerals yearbook notes that, 'in 2016, PGMs were recycled from three main sources — catalytic converters, electronics, and jewellery. Globally, more than 121,000 kg of secondary Pgms was recovered, accounting for approximately 23% of the global supply' (USGS, 2016: 57.1). Of that amount, 'about 77,400 kg [was] palladium' with 13,600 kg recovered from electronic devices (USGS, 2016: 57.2).

## ELEMENT NO. 47: SILVER (AG)

Technically considered part of the Pgms, silver is much more widely known than other Pgms and has a history of use almost as long as humanity itself. One of the precious, non-reactive "noble metals" silver is discoverable in an already mostly-pure form in the environment, as veins, chunks, nuggets, etc. with many similar properties to both copper and gold (all sharing a column on the periodic table), being a good conductor, quite soft and malleable. More abundant than gold, silver usually trades around 10 to 20 USD per ounce, while a single ounce of gold can fetch thousands. As a result silver can be used economically in a much wider range of applications than gold, though it remains more expensive than copper.

When one thinks of silver, one might first think of jewelry, however the industrial applications of silver are substantial, using one quarter of the world's silver production, mainly for use in lead-free solder (Manhart et al., 2016: 20). A critical part of the modern electronics industry, solder is an alloy of tin (see element no. 50) and other metals that allow it to be melted at a relatively low temperature, used to join electrical components or connections in a more permanent way, as once it cools it quickly re-solidifies. Historically, solder began as a mix of tin and lead (see element no.82) giving it a low enough melting point to be used as solder. The European Union's directives on Waste Electrical and Electronic Equipment (WEE) and Restriction of Hazardous Substances (RoHS) came into force in 2006 banning the use of lead in solder, given the toxic nature of lead (a heavy metal). Solder itself plays no small part in enabling the high-throughput of mass-produced electronics manufacturing, and it would be almost impossible without the ability to durably join transistors, capacitors, wires and input/output devices to the rest of circuit. The reversibility of the connection solder makes also help allow for easy repair or replacement of parts, as when heated enough it returns to its viscous liquid form, allowing the part to be swapped or removed. Modern lead-free solder is almost certainly the main reason for the presence of silver in the PS4 APU sample, with the sample obtained by directly cutting it from the PCB it was attached to, potentially including some of the solder attaching the

APU to the main board. Furthermore, the presence of silver adds weight to the previous suggestion of some use of palladium-silver thick films in the PS4 APU.

For a metal with such a long history, it should come as no surprise that its environmental harms have an equally long story. Silver mining dates back at least five thousand years and frequently involved the use of the toxic heavy metal mercury until the development of modern methods involving cyanide. Silver (together with gold) can be refined using the "patio" method, a process mostly unchanged since Roman times. Lacerda (1997) describes it as follows:

> The "Patio" process consists of spreading silver and gold powdered ore over large, paved flat surfaces and mixing it with salt brine and a mixture of Cu (copper) and Fe (iron) pyrites and elemental Hg (mercury). Workmen blended the mixture with hoes and rakes and let it react for days to weeks, depending on the weather, for amalgamation. After removing the amalgam, Au (silver), Ag (gold) and Hg (mercury) were recovered through roasting. (Lacerda, 1997: 211)

The toxicity of mercury is well documented and understood, with its vapour highly poisonous to humans and animals. Mercury readily accumulates in the environment typically as methylmercury (a carbon-hydrogen-mercury mixture) which is particularly impactful on aquatic life, and which bioaccumulates, eventually making its way up the food chain into humans (Lacerda, 1997: 209). And while the patio method has largely been replaced in the west by the cyanidication process (more on which in a moment) according to Lacerda (1997: 213) the increase in gold prices during the 1970s resulted in a new gold rush in parts of the developing world which often returned to the ancient, dangerous and environmentally destructive mercury based method. One quirk of this explosion of gold and silver mining is that it provided researchers with the possibility of comprehensively estimating the historical and contemporary mercury emissions associated with mining for both these precious metals, with Lacerda offering the worrying conclusion that 'a tentative estimate of past [mercury] emissions to the environment may reach an astonishing amount of over 260,000 t over the last 400 yr.' (Lacerda, 1997: 213). The World Health Organization notes the following devastating toxic consequences of exposure to various forms of mercury:

Elemental and methylmercury are toxic to the central and peripheral nervous systems. The inhalation of mercury vapour can produce harmful effects on the nervous, digestive and immune systems, lungs and kidneys, and may be fatal. (WHO, 2017)

For a sense of how little it takes to impact a person, whether a worker roasting via the patio method, or just as commonly a person who survives on a diet of fish highly contaminated with mercury, chronic exposure can be detected even from microgram doses:

Mild, subclinical signs of central nervous system toxicity can be seen in workers exposed to an elemental mercury level in the air of 20 $\mu g/m^3$ or more for several years. (WHO, 2017)

With the knowledge that as much as 260,000 tons of mercury may have been released into the environment over the past 400 years, it is easy to see how much human (and animal) suffering is indirectly attributable to gold and silver mining. Modern mercury-free methods thankfully involve the leeching of gold and silver by cyanide solutions and avoid the use of mercury altogether. The use of cyanidation however still produces so-called "cyanidation wastes" with Lottermoser (2007: 183) noting that 'at modern gold mining operations, cyanidation wastes occur in the form of heap leach residues, tailings, and spent process waters' (2007: 183). While the cyanide process waters can be recycled, with more cyanide added to be reused in future leaching, at various points in the cycle these waters need storage, usually in large tailing ponds. In these ponds the cyanide levels naturally reduce over time, turning into less harmful compounds and molecules as UV light helps break it down, avoiding some of the accumulation problems associated with elemental and methyl-mercury (Lottermoser, 2007). While there have been cases of cyanide spillages, none have been known to cause human deaths, however in enough concentration it can poison fish, animals and humans. Lottermoser (2007: 192) notes that despite its reputation as a poison-pill for spies, 'cyanide is not toxic to plants, and a major cyanide spill in the Kyrgyz Republic had no impact on plant life.' For these reasons, the use of cyanide is greatly preferable to mercury, with cyanide at least an organic, naturally occurring molecule present in small amounts in plants and animals, even 'an important component of vitamin B-12' (Lottermoser, 2007: 183).

Clearly there are better and worse options for silver and gold mining, and it remains an important global challenge—one which device manufacturers have a part to play in by ensuring that mercury based methods are replaced. Unlike mercury, which as an atomic *element* that cannot be further broken down, instead persisting in the environment, cyanide is (at the risk of oversimplifying) a molecule made out of some of the most common elements already in the environment: carbon and nitrogen, sometimes dissolved in water, sometimes in salt forms, and contains no heavy metals. Because of this, despite its reputation for delivering a quick death for the captured spy, cyanide is much easier for living organisms to break down and less of an environmental hazard long-term.

One final consequence of silver and gold mining that might be less obvious is found in their effect on humans and human societies. Silver mining has been around for thousands of years, with some of the oldest human civilizations organized around the extraction of gold and silver, extremely dangerous work, often done by slaves. David Graeber (2011) makes the argument in *Debt: The First 5000 Years* that gold and silver often turn up in the anthropological record, enabling something akin to the depersonalization of human relations. As a stable container of "value" they allow societies to develop exchange systems without the reciprocal relationships of social credit and debt, which Graeber instead describes as the earliest 'human economies', preceding the invention of coinage. Precious metals, wherever they emerge in world history, according to Graeber (2011: 213) 'served the same role as the contemporary drug dealers suitcase of unmarked bills: an object without a history, valuable because one knows it will be accepted in exchange for other goods just about anywhere, no questions asked.' The result, he claims, is that 'credit systems tend to dominate in periods of relative social peace... in periods characterized by war and plunder they tend to be replaced by precious metals' (Graeber, 2011: 213). The durability and conductivity of gold and silver forms the core of their use-value for the electronics industry. But the same stability—and the fungibility of equal weights of the metal—gives it power as a store of value in times of conflict and war. This interchangeability makes the provenance of any given piece of gold or silver present in the PS4 almost impossible to trace. On balance, they are almost certain to have been mined from a modern mine, rather than melted down from ancient sources, recycled from gold or silver coin and jewelry, however it's utterly impossible to be sure. All the gold and silver in the world ever mined, from those mined thousands of years ago in conditions of abject squalor via harmful mercury-based methods still exist in the world today. Like other elements in this list, the connection of any one particular atomic element, the components it is associated with, and its direct responsibility for particular harms is difficult, if not impossible to prove. Nevertheless we should not overlook the global tragedy that is relationships of violence and exploitation, human dominance of the natural world, and the harms from these processes, all of which haunt these commodities, lurking in the history and supply chains which can be made visible, like UV light on dollar bills that show traces of the grime and bodily fluids that have touched them.

## ELEMENT NO. 48: CADMIUM (CD)

Despite being adjacent to non-reactive silver, cadmium is a highly-reactive, toxic heavy metal. It is located in the same column and closely related to mercury (see element no. 47 silver for a discussion of mercury). The application of cadmium, according to the USGS, is mainly in 'alloys, coatings, [nickel-cadmium] batteries, pigments, and plastic stabilizers' (USGS, 2019: 40). Battery production for nickel-cadmium (NiCad) batteries accounts for most of the world's consumption, but it also has a use in the production of certain types of solar cells (USGS, 2019: 41). One of the oldest battery chemistries, the NiCad was first discovered in the late 1890s and remained a popular choice as a rechargeable power source until its replacement with the less toxic nickel-metal hydride (NiMH) and the eventual development of lithium-ion (Li-ion) batteries. The European Union has banned the use of NiCad batteries in devices, but in other countries (like Australia) they can still be found in things like my electric toothbrushe.

While we have already mentioned that batteries are not part of the PS4 APU, there are however some rather rare and exotic semiconductor materials like cadmium telluride (tellurium element no. 52—also detected but not in large quantities), cadmium selenide, and others. These are more typically used in photovoltaics, optical and infrared sensors, so unlikely to be useful to the PS4 APU. Cadmium oxide, however, can be used as an n-type semiconductor which seems the most likely culprit, potentially being used to construct some part of the billions of transistors that make up the graphics processing and other logic functions of the APU (Kasap & Capper, 2017).

The European Union's Restriction of Hazardous Substances (RoHS) directive limits the use of several substances including a number raised already in this chapter—lead, mercury, and hexavalent chromium—as well as cadmium. The use of cadmium in particular is restricted to the strictest level of less than 100 parts per million, with lead, mercury and hexavalent chromium limited only to less than 1000 parts per million (ROHS guide, 2021). Products manufactured after 2011 that carry the 'CE' logo on them must conform to these EU RoHS standards and 'have been assessed to meet high safety, health, and environmental protection requirements' (European Commission, 2021). The outer plastic shell of the PS4 does not seem to carry this logo, however the icon does appear on the plastic shell of the power-supply. It is unclear whether the CE on this part refers to the whole device or just the PSU, and the ICP-MS test as conducted was unable to tell us concentrations or quantities of cadmium detected.

Cadmium is indisputably toxic, with the EPA describing its effects on aquatic life in no uncertain terms. Deadly to aquatic life, 'chronic exposure leads to adverse effects on growth, reproduction, immune and endocrine systems, development and behavior in aquatic organisms' (EPA, 2016). Rao et al. (2017: 221) provides the following summary of similarly devastating effects in humans:

> Due to its long half-life in the body, cadmium can easily be accumulated in amounts that cause symptoms of poisoning. Cadmium shows a danger of cumulative effects in the environment due to its acute and chronic toxicity. Acute exposure to cadmium fumes causes flulike symptoms of weakness, fever, headache, chills, sweating, and muscular pain. The primary health risks of long-term exposure are lung cancer and kidney damage. Cadmium is also believed to cause pulmonary emphysema and bone disease (osteomalacia and osteoporosis).

The presence of this toxic substance inside the APU presents a solid case for top priority elimination. However we might reasonably suspect that levels are under the limits specified by the RoHS (more testing is required to verify this) in any case at the very least it is an important reminder of the appropriate handling and disposal of e-waste, and if alternatives or substitutions are available that they should be considered. Recycling of PS4s to avoid even these small amounts of cadmium being released into the environment ought to be strongly encouraged as these devices rapidly approach their use-by date.

## ELEMENT NO. 49: INDIUM (IN)

Indium is a little-known and rarely discussed element, not the kind of thing that comes up in ordinary conversation. This is despite being almost ubiquitous, essential in LCD screens, and enabling the capacitive touch screens of smartphones—the compound material indium tin oxide (ITO) creating a thin conductive layer that remains transparent. This is so essential that the USGS notes that 'production of indium tin oxide (ITO) continued to account for most of global indium consumption' (USGS, 2019: 78). Other uses for indium in the electronics industry include solders made of indium and lead which are 'used to inhibit the leaching of gold components in electronic apparatus [as well as] Indium-silver alloys or pure indium foil... used as thermal interface materials in electronics (a substance used to seal a heat-generating surface to a heat sink, filling microscopic air voids to allow for effective heat transfer)' (USGS, 2018: 35.1). As conducting heat away from the PS4 APU is critical to its operation, there is a reasonable chance that indium-silver compound or similar may be used in the APU for this purpose. The newer PlayStation 5 console makes use of a similarly exotic 'liquid metal' thermal compound instead of classic thermal paste. The fact sheet of a similar (though perhaps not identical) liquid metal thermal paste lists its composition as an 'alloy of the metal components gallium, indium, rhodium, silver, zinc and stannous, bismuth; suspended in a graphite-copper matrix' (CoolLaboratory, 2010: 1). There are also other semiconductor uses for indium which may explain its presence in the APU (USGS, 2018: 35.1) and elsewhere there are applications of indium in photovoltaic cells.

In terms of its production, according to the USGS 'indium is most commonly recovered from the zinc-sulfide ore mineral sphalerite' (USGS, 2019: 79). Pradhan et al. (2018: 167–8) explain that the zinc refining process 'enriches the indium content in its solid by-product which is further used as the primary mineral resource of indium.' The dual production of zinc and indium somewhat complicates estimates of emissions (but see the figures for zinc on its own in element no. 30). Takahashi et al. (2009: 891) in a study of potential LCD screen recycling of indium concluded that: 'the indium content in zinc ores varies from 10 to 20 g/t, which is much smaller than its content in the LCD of discarded cellular phones after all the parts made of organic materials are removed.' This suggests that, with sufficiently developed recycling, old electronic devices could become a significant source of indium in the future, almost certainly reducing emissions and the amount of material ending up in landfill.

In terms of environmental and health consequences, details are hard to come by. According to Guha et al. (2017), workers may become exposed to indium tin oxide 'during ITO production and processing, or during elemental indium recycling' and though the primary toxicological data available is from animal studies, it is classified as possibly carcinogenic (Guha et al., 2017: 582).

## ELEMENT NO. 50: TIN (SN)

Our wide cultural familiarity with tin, turning up in phrases like 'tin ear' and so on, is perhaps more to do with its long history of use than its abundance, having being known since the bronze age, the metal is actually relatively scarce. The USGS provides a figure for its abundance in the earth's crust of only 'about 2 parts per million (ppm), compared with 94 ppm for zinc, 63 ppm for copper, and 12 ppm for lead.' A breakdown of the uses of the metal in the United States is also provided by the USGS (2019) with its use in solder (see also element no.47 silver) accounting for 14% of production, however Yang et al. (2018: 1352) place this figure even higher, citing a figure of 'approximately 44% of global refined tin... used as solder in the electronics industry' rising to 61% in China. Most likely then tin is present primarily as part of the ball-grid array of tiny soldered connections that attaches the dozens of communication pins on the underside of the APU to the rest of the circuit board or from other soldered connections.

The US EPA has insufficient data to provide recommendations on exposure limits or details about the toxicity of tin for humans or animals, with studies of former tin mines repurposed for other activities tending to focus on the secondary effects of tin mining. These include the exposure of heavy metals like uranium and thorium in former Nigerian tin mines, (Ibeanu, 2003) lead, chromium and arsenic in Malaysian mines, (Yusof et al., 2001; Alshaebi et al., 2009) and issues of soil restoration around tin mines in Indonesia (Agus et al., 2017). Yang et al. (2018) note that the solid-waste burden of tin mining is substantial (about 0.8 times the resulting ore, almost half of the removed rock being useless 'waste'), providing a range of energy intensities for different forms of tin mining from 11.43 MJ up to 66.24 MJ of energy per ton. This represents a GER (for 1 kg of tin ore) that is around 4 orders of magnitude smaller than the most energy intense common metal Titanium (element no.22), placing tin lower on our list of concerns.

## ELEMENT NO. 56: BARIUM (BA)

Element no. 56 is Barium, a heavy, dense metal, more commonly used as a weighting agent, 90–95% of the world's production being used in the petroleum industry as part of drilling and the creation of oil wells. An explanation for its presence in the PS4 APU sample is therefore not immediately evident. The most likely case seems to be due to its use in capacitors, with the Springer handbook of electronic and photonic materials noting that certain 'BaTiO$^3$-based ceramics are widely used in ceramic capacitors and form the basis of an industry worth billions of dollars annually' (Kasap & Capper, 2017: 595). It is possible that some miniaturized capacitors are part of the PS4 APU's logic circuits, however this is simply speculation. Acosta et al. (2017) notes that uses of barium titanate have emerged in the years following the implementation of the EU's RoHS, with its potential as a replacement for lead-based components, like lead zirconate titanate (or PZT—see element no. 22 for a discussion) (Acosta et al., 2017: 2).

Toxic and non-essential to life, barium is an undesirable heavy metal, not normally present in any sort of concentrated level in most environments. As Rao (2017: 220) notes:

> being highly unstable in the pure form, it forms poisonous oxides when in contact with air. Short-term exposure to barium could lead to brain swelling, muscle weakness, and damage to the heart, liver, and spleen. Animal studies reveal increased blood pressure and changes in the heart from ingesting barium over a long period of time. The long-term effects of chronic barium exposure to human beings are still not known due to lack of data.

Despite its toxicity, however, and largely due to its reactivity in air such that it almost always is encountered in compound or oxide form, according to the USGS its 'limited mobility in the environment... poses minimal risk to human or ecosystem health' (USGS, 2017: D.1). One study of rice grown in areas adjacent to barium mining in several regions of China found 'extremely high Ba concentrations existed in paddy soils, especially near the Ba mining area as well as Ba salt plants' (Lu et al., 2019: 147). The authors of the report go on to state that 'rice can accumulate high Ba concentrations' however in all the tested samples, exposure levels were still below US EPA recommended limits (Lu et al., 2019: 147).

There seems to be little research or data available on the energy requirements of barium mining and refining. It can be mined via either open-cut
pit or underground. Barium mining in South Australia typically occurs via
the underground method (Dept of Energy and Mining, Government of
South Australia, 2021) which is typically more energy intensive than open-
cut mines. I could find no estimates of a GER or GWP for barium mining,
however in 2018 the US only produced around 480,000 tons of the substance (USGS, 2019: 28).

## ELEMENT NO. 79: GOLD (AU)

As one of the least reactive metals, gold is prized for its purity, rarity, and
other useful physical properties critical to industrial applications. Used by
humans for thousands of years, like its close relative silver (in the same
column of the period table), it is stable and generally non-reactive. In
electronics, it is often used as plating on connectors as gold does not tar-
nish, and it can also be used on its own as a wire or paste. There are other
niche uses for gold, like for example as part of nanoscale memory storage
in a highly specific non-silicon based flash memory arrangement (Kasap &
Capper, 2017: 1273). The handbook again provides a nice summary of its
characteristics:

> Gold pastes have a high conductivity and are mainly used in applications
> where high reliability is required. Gold is a particularly good material for
> wire bonding pads, although it has relatively poor solderability. Gold is a
> precious material and hence very expensive; it is therefore not used for gen-
> eral purpose applications and is limited to those areas that can justify the
> higher costs. (Kasap & Capper, 2017: 711)

Mentions elsewhere throughout the volume are limited to its role as a
bonding material or a connector, and we can tentatively conclude that its
role in the PS4 APU is likely to be related to one of these properties.

For discussion of the environmental harms of gold, see element no. 47
(silver) which covers much of the same mining and refining process. Across
the fifteen years from 1991 to 2006, Mudd (2007: 49, table 2) provides
the following averages for the production of one kilogram of gold: 691 kL
of water consumed, 11.5 tonnes of $CO^2$ equivalent emissions, 143 GJ of
energy used, and 141 kg of cyanide consumed. As these are averages, significant outliers exist both in time and across the different locations that

gold mining occurs. Suffice to say the resource intensity of gold mining is substantial. As Mudd (2007: 53) summarizes the situation, which mirrors the earlier discussion of PGMs staggering energy intensity (see element no. 46 palladium): 'although the high mass ratio of $CO^2$ to gold is due to the relatively small mass of gold produced, the primary function of gold for jewellery leads to a major ethical and social issue in terms of accounting for the greenhouse costs.' The same must surely hold for its role in electronics manufacturing, despite the lower consumption of gold on a per-device basis, the scale of production means this will still add up.

## ELEMENT NO. 82: LEAD (PB)

Another widely known metal used for thousands of years, lead is a toxic heavy metal known to cause serious health issues in those exposed to it. As Rao (2017: 222) concisely summaries,

> Short-term exposure to high levels of lead can cause vomiting, diarrhea, convulsions, coma, or even death. Other symptoms are appetite loss, abdominal pain, constipation, fatigue, sleeplessness, irritability, and headache. Continued excessive exposure, as in an industrial setting, can affect the kidneys. It is particularly dangerous for young children because it can damage nervous connections and cause blood and brain disorders.

Exposure most commonly comes though some form of ingestion, such as through drinking contaminated water, as in the city of Flint, Michigan where the town's water supply became contaminated by corroded lead pipes.

The USGS (2019: 94) estimates that 'the lead-acid battery industry accounted for more than 85% of reported U.S. lead consumption during 2018.' In the electronics industry, lead has previously been used as a component in solder (see element no.50 tin for a discussion) however its use here (as elsewhere) have mostly been phased out following the EU's Restriction on Hazardous Substances (RoHS). In response, 'the electronics industry has moved toward lead-free solders and flat-panel displays that do not require lead shielding' (USGS, 2019: 95). Other exotic uses of lead in electronics include various compounds important for thermal imaging (lead–antimony–silver–telluride, lead sulphide, and others) and piezoelectric sensors that detect pressure, vibration, and so on, described as having 'enormous technological and commercial importance' by the Springer Handbook (Kasap & Capper, 2017: 595). The Handbook also notes that ferroelectric materials like lead-zironate-titanate (PZT—see element no. 22)

> have been utilized for some time as dielectrics for non-volatile memory elements based upon the ferroelectric effect. In storage capacitors incorporating such materials, the polarization state of the ferroelectric is preserved (once poled) without the presence of an electric field. This state is then sensed by associated circuitry and can therefore be used as a memory device. (Kasap & Capper, 2017: 638)

Whether the PS4 contains this particular type of memory inside the APU, or whether it is some other explanation like lead-based solder at levels below those permitted by the RoHS is unclear from testing.

Environmentally, the risks of lead in electronic waste streams in particular, are significant, with Rao et al. (2017: 217) noting that 'significant amounts of lead are dissolved from broken lead-containing glass, such as the cone glass of cathode ray tubes (CRTs), which gets mixed with acid waters; this is a common occurrence in landfills.' One positive sign is that recycling of lead is increasingly accounting for lead required by industry. As Wilburn (2015: 1) explains in a study of the lead recycling industry in the United States: 'Since 1995, domestic production of lead has increasingly shifted from primary mining and smelting to the recovery of lead-bearing scrap by the secondary lead industry, which accounted for 91 percent of U.S. lead production in 2012.' Mostly, however, this seems to involve the recycling of lead-acid batteries, and reports from as recently as 2012 describe stockpiles of millions of pounds of lead-containing glass left over from old CRT monitors (Kyle, 2012). Hong et al. (2017: 909), performed a life cycle assessment of primary and secondary lead refining in China for the year 2013, providing some sobering estimates: 'approximately 5.61 Mt $CO^2$ eq.... 0.4 kt lead, 18.4 kt sulfur dioxide, 15.6 kt nitrogen oxide, and 6.4 kt particulate emissions...were released from the lead industry in China.' Clearly the presence of *any* lead in substantial quantities in the PS4 APU is worrying, and of some concern. The finding of it here should be followed by the asking of some uncomfortable questions of manufacturers.

## ELEMENT NO. 83: BISMUTH (BI)

The final element in our list is bismuth. The only heavy metal non-toxic to humans, the bulk of industrial applications of bismuth are 'in cosmetic, industrial, laboratory, and pharmaceutical applications' (USGS, 2019: 34). It is the main ingredient in Pepto-Bismol, for instance, the over-the-counter stomach antacid. The USGS (2019: 34) also notes that 'bismuth-tellurium-oxide alloy film paste is used in the manufacture of semiconductor devices' with ferroelectric and semiconductor applications which may account for its appearance in the PS4 APU sample (see element no. 82 Lead for mention of ferroelectrics) (Kasap & Capper, 2017: 286, 543).

Despite its non-toxicity, production of Bismuth happens 'most often [as] a byproduct of processing lead ores' (USGS, 2019: 35) thereby associating its production with that of lead and its substantial environmental harms. In 2016 'world refinery production of bismuth was estimated to be 17,100 t' (USGS 2016: 12.1) so it is not a huge volume of the mineral being produced. It's production as a by-product of other forms of mining once again complicates emissions estimates, which I have been unable to find. Though it has a reputation as a relatively harmless element, owing to benign uses in human treatments and medicines, Filella (2010) has noted that 'the increased use of bismuth because of its 'green' reputation requires a better understanding of its environmental behaviour' (Filella, 2010). After conducting a review of the literature, Filella (2010) concludes that it is not possible to establish 'a sound estimation of bismuth concentrations in seawater. Nor is it possible to identify a range of probable concentration values for freshwaters which are not heavily polluted' and that 'regrettably, the unsatisfactory situation concerning bismuth is not unique; our current understanding of the chemical elements' environmental behaviour being based too often on redundant studies of a limited number of elements.' This is a reminder to not assume that lesser, or even simply *less well known* elements means that there are no harms involved. It also suggests that quite a lot more research is still needed to fill out the picture of what the ecological effects are of many of these critically important materials inside our high tech devices.

## CONCLUSIONS

This chapter has followed the huge range of elements detected in the PS4 APU sample by ICP-MS analysis, tracing them back to their sources in the earth's crust. I have attempted sketch out how they are transformed into raw material feedstock for modern semiconductor manufacturing, paying attention to the energy used, the nature and scale of mining impacts, touching on elements of the industrial process as diverse as lithographic nanometer scale circuit construction and solid-waste burdens from rock removal at sites all around the world. The picture this presents is only a partial one, an initial foray, followed I hope by much more pointed and targeted investigation into the particular processes and harms that are the consequences of modern digital device manufacturing. The demand for these devices is growing at an incredible rate, and must not be overlooked in our response to the demands of the climate crisis. What we now have, at least partially, is a sense of the emissions embodied in these mass-produced devices. The periodic table of torture shows that there are no innocent devices, though some may be more or less harmful, and the challenges of producing a hardware device like a new gaming console—the eventual PlayStation 6 or the next Xbox—are substantial, and the efforts to decarbonize, and reduce their harms to the planet is urgent. The periodic table of torture also gives us a sense of where we might wish to prioritize our efforts as well. From the more toxic heavy metals—like lead, cadmium, and barium—alternatives, or different manufacturing methods may need to be developed to reduce our dependence on them. Similarly, a much greater end-of-life focus on the recycling of devices—perhaps a manufacturer buy-back scheme—could prove useful in avoiding long-lasting damage to wherever this waste would otherwise end up. This will no doubt be challenges to proposing and implementing regulation and political will is required. Greater recycling—and the identification of the *right methods of recycling* for each particular device, whether manual crushing to recapture copper for instance, or PGM targeting methods, should be determined and disseminated. Greater use of recycled metals for those with high GWP and GER for virgin metal—like aluminium and titanium—could help push down the overall emissions burden.

The periodic table of torture above all else shows just how much work remains ahead of us, and just how big of a problem gaming hardware is. The carbon neutral games industry cannot continue to produce new devices, new consoles, with the same frequency as in the past—it is simply unsustainable to continue to use the same materials and processes as it currently does. Microsoft may have been able to claim that it was able to manufacture a certain number of Xbox One consoles in a carbon neutral way, as we saw in Chap. 4. But as the analysis in this chapter shows, even neutralizing and offsetting carbon emissions for the manufacture of these devices is unlikely to be adequate to the task of addressing all current and future harms associated with the materials that have gone into the device. What happens to these devices when they are thrown out, replaced with a newer generation of consoles, remains just as critical to our efforts to forestall the worst ecological disaster this planet has seen for millennia.

The modern games industry is a bit like the city of Omelas in Ursula Le Guin's short story. Omelas is a perfect city in which all the art, joy and life is contingent on the suffering of a single poor, wretched individual, locked away in misery and suffering:

> They all know it is there, all the people of Omelas. Some of them have come to see it, others are content merely to know it is there. They all know that it has to be there. Some of them understand why, and some do not, but they all understand that their happiness, the beauty of their city, the tenderness of their friendships, the health of their children, the wisdom of their scholars, the skill of their makers, even the abundance of their harvest and the kindly weathers of their skies, depend wholly on this child's abominable misery. (Le Guin, 1975)

Like all capitalist production, the suffering that the games industry relies on is not even this neatly confined, instead it is spread across a host of people, plants, animals, environments. We too are more than ever before aware that our enjoyment, our 'homes' in the glittering city of gaming, rests upon a mountain of e-waste, on tons of carbon dioxide and other noxious emissions, on the transformation of the very earth itself into intricate devices, and the devastating exacerbation of the hot-house effect on the planet's atmosphere. The central subject of Le Guin's short story is not really the city itself, however, but those few who can't stand this deeply immoral arrangement:

> They leave Omelas, they walk ahead into the darkness, and they do not come back. The place they go towards is a place even less imaginable to most of us than the city of happiness. I cannot describe it at all. It is possible that it does not exist. But they seem to know where they are going, the ones who walk away from Omelas.

Opting out of the games industry, much like walking away from the modern world, is not really possible, nor a solution. We may not have a complete picture of what must be done but if we believe games are worth saving, along with the planet, then this is now our obligation: to extend our imagination and concern as far as they will go, to bring into being the digital games industry that comes after climate change. We are aware of the debts we are racking up and that we have to do something about it. Rather than walk off into the wilderness (there is, after all, nowhere else to go) we need to develop in ourselves a capacity to imagine a better future, the anti-Omelas. The modern, sustainable, utterly carbon-neutral games industry. A future for games that contributes to restoring the world, nourishing and sustaining the very life and country it is made on, rather than destroying it.

The challenge for readers is now to take up the task of thinking how we might get from here to there—to the sustainable and carbon neutral games industry that we sorely need. Creative solutions will be necessary, and they are needed desperately soon. I believe that the carbon neutral games industry is possible, though it might not look exactly how we expect, or even how we might want it to today. Answering how we might get there exactly is beyond the scope of this chapter, and even to an extent beyond the scope of this book. In the next chapter I offer some of my own very modest suggestions in the hope that they might inspire more, much better ideas from others.

## REFERENCES

Acosta, M., Novak, N., Rojas, V., Patel, S., Vaish, R., Koruza, J., Rossetti, G. A., Jr., & Rödel, J. (2017). BaTiO3-based piezoelectrics: Fundamentals, current status, and perspectives. *Applied Physics Reviews, 4*(4), 041305.

Agus, C., Wulandari, D., Primananda, E., Hendryan, A., & Harianja, V. (2017). The role of soil amendment on tropical post tin mining area in Bangka island Indonesia for dignified and sustainable environment and life. *IOP Conference Series: Earth and Environmental Science, 83*(1). IOP Publishing.

Alshaebi, F. Y., Yaacob, W. Z. W., Samsudin, A. R., & Alsabahi, E. (2009). Risk assessment at abandoned tin mine in Sungai Lembing, Pahang, Malaysia. *Electronic Journal of Geotechnical Engineering, 14*, 1–9.

AmCham Chile. (n.d.). *Chile's Mining Industry.* https://www.amchamchile.cl/UserFiles/File/Mining%20Industry.pdf. Accessed 1 Apr 2021.

American Elements. (2020). *Zinc (Zn) | American elements.* https://www.americanelements.com/zn.html. Accessed 1 Apr 2021.

Andrae, A. S. G., & Edler, T. (2015). On global electricity usage of communication technology: Trends to 2030. *Challenges, 6*(1), 117–157.

Aslan, J. (2020). *Climate change implications of gaming products and services.* PhD dissertation, University of Surrey.

Australian Aluminium Council. (2020). *Bauxite mining.* https://aluminium.org.au/interactive-flowchart/bauxite-mining-chart/. Accessed 1 Apr, 2021.

Bartos, S. C. (2002). *Update on EPA's magnesium industry partnership for climate protection.* Presented at the 131st TMS Annual Meeting, February 17–21, Seattle, Washington.

Bratton, B. H. (2016). *The stack: On software and sovereignty.* MIT press.

Chen, H.-W. (2006). Gallium, indium, and arsenic pollution of groundwater from a semiconductor manufacturing area of Taiwan. *Bulletin of Environmental Contamination and Toxicology, 77*(2), 289–296.

CoolLaboratory. (2010). EG-SAFETY DATA SHEET (EG Nr. 1907/2006). http://www.coollaboratory.com/pdf/safetydatasheet_liquid_ultra_englisch.pdf

Cubitt, S. (2016). *Finite media.* Duke University Press.

Das, S. (2012). Achieving carbon neutrality in the global aluminum industry. *JOM, 64*(2), 285–290.

Dept of Energy and Mining, Government of South Australia. (2021). *Barite.* https://energymining.sa.gov.au/minerals/mineral_commodities/barite. Accessed 2 Apr 2021.

EPA. (1998). Chromium (VI); CASRN 18540-29-9. *Integrated Risk Information System (IRIS) chemical assessment summary.* U.S. Environmental Protection Agency, Washington, DC. https://cfpub.epa.gov/ncea/iris/iris_documents/documents/subst/0144_summary.pdf. Accessed online 13 Jan 2020.

EPA. (2005). *Toxicological review of zinc and compounds.* U.S. Environmental Protection Agency, Washington, DC.

EPA. (2008). *Lead and copper rule: Quick reference guide.* US Environmental Protection Agency Office of Water. https://www.epa.gov/dwreginfo/lead-and-copper-rule

EPA. (2016, March). *Aquatic life ambient water quality criteria Update for Cadmium – 2016.* US Environmental Protection Agency Office of Water. http://www.epa.gov/wqc/aquatic-life-criteria-cadmium. Accessed 2 Jan 2020.

EPA. (2018). *Toxicological review of hexavalent chromium.* U.S. Environmental Protection Agency, Washington, DC.

EPA. (2020). *Aquatic life criteria – Copper.* U.S. Environmental Protection Agency, Washington, DC. https://www.epa.gov/wqc/aquatic-life-criteria-copper. Accessed 1 Apr 2021.

European Commission. 2021. *Internal market, industry, entrepreneurship and SMEs | CE marking.* https://ec.europa.eu/growth/single-market/ce-marking/. Accessed 2 Apr 2021.

Filella, M. (2010). How reliable are environmental data on 'orphan' elements? The case of bismuth concentrations in surface waters. *Journal of Environmental Monitoring, 12*(1), 90–109.

FranceTVInfo. (2013, February 17). Le long combat contre la pollution de la Méditerranée par la Montedison. *FranceTVInfo.* https://france3-regions. francetvinfo.fr/corse/2013/02/17/le-long-combat-contre-la-pollution-de-la-mediterranee-par-la-montedison-201739.html. Accessed 1 Apr 2021.

Fuchs, C. (2008). The implications of new information and communication technologies for sustainability. *Environment, Development and Sustainability, 10*(3), 291–309.

Garnaut, R. (2019). *Superpower*. Black Inc.

Glaister, B. J., & Mudd, G. M. (2010). The environmental costs of platinum–PGM mining and sustainability: Is the glass half-full or half-empty? *Minerals Engineering, 23*(5), 438–450.

Gordon, L. (2019, December 5). The environmental impact of a PlayStation 4. *The Verge*. https://www.theverge.com/2019/12/5/20985330/ps4-sony-playstation-environmental-impact-carbon-footprint-manufacturing-25-anniversary

Graeber, D. (2011). *Debt: The first five thousand years*. Melville House.

Grossman, E. (2006). *High tech trash: Digital devices, hidden toxics, and human health*. Island Press.

Guertin, J., Jacobs, J. A., & Avakian, C. P. (Eds.). (2004). *Chromium (VI) handbook*. CRC Press.

Guins, R. (2014). *Game after: A cultural study of video game afterlife*. MIT Press.

Guha, N., Loomis, D., Guyton, K. Z., Grosse, Y., El Ghissassi, F., Bouvard, V., Benbrahim-Tallaa, L., Vilahur, N., Muller, K., & Straif, K. (2017). Carcinogenicity of welding, molybdenum trioxide, and indium tin oxide. *The Lancet Oncology, 18*(5), 581–582.

Gura, D. (2010, October 5). Toxic red sludge spill from Hungarian Aluminum Plant 'An Ecological Disaster'. *NPR*. https://www.npr.org/sections/thetwo-way/2010/10/05/130351938/red-sludge-from-hungarian-aluminum-plant-spill-an-ecological-disaster. Accessed 1 Apr 2021.

Haque, N., & Norgate, T. (2014). The greenhouse gas footprint of in-situ leaching of uranium, gold and copper in Australia. *Journal of Cleaner Production, 84*, 382–390.

Hong, J., Zhaohe, Y., Shi, W., Hong, J., Qi, C., & Ye, L. (2017). Life cycle environmental and economic assessment of lead refining in China. *The International Journal of Life Cycle Assessment, 22*(6), 909–918.

Ibeanu, I. G. E. (2003). Tin mining and processing in Nigeria: Cause for concern? *Journal of Environmental Radioactivity, 64*(1), 59–66.

Jondreville, C., Revy, P. S., & Dourmad, J.-Y. (2003). Dietary means to better control the environmental impact of copper and zinc by pigs from weaning to slaughter. *Livestock Production Science, 84*(2), 147–156.

Kasap, S., & Capper, P. (Eds.). (2017). *Springer handbook of electronic and photonic materials.* Springer.

Kittler, F. (1995). There is no software. *ctheory.net*, 10–18.

Kittipongvises, S., Chavalparit, O., & Sutthirat, C. (2016). Greenhouse gases and energy intensity of granite rock mining operations in Thailand: A case of industrial rock-construction. *Environmental & Climate Technologies, 18*(1).

Kwon, J. Y., Son, K. S., Jung, J. S., Kim, T. S., Ryu, M. K., Park, K. B., Yoo, B. W., et al. (2008). Bottom-gate gallium indium zinc oxide thin-film transistor array for high-resolution AMOLED display. *IEEE Electron Device Letters, 29*(12), 1309–1311.

Kyle, B. (2012). Recyclers stockpiling millions of pounds of toxic glass from CRT TVs and monitors. *Electronics Take Back Coalition.* http://www.electronic-stakeback.com/2012/11/15/recyclers-stockpiling-millions-of-pounds-of-toxic-glass-from-crt-tvs-and-monitors/. Accessed 29 July 2021.

Lacerda, L. D. (1997). Global mercury emissions from gold and silver mining. *Water, Air, and Soil Pollution, 97*(3), 209–221.

Lavigne, M. J., & Michaud, M. J. (2001). Geology of North American Palladium Ltd.'s Roby zone deposit, Lac des Iles. *Exploration and Mining Geology, 10*(1–2), 1–17.

Le Guin, U. K. (1975). "The ones who walk away from Omelas" in Silverberg, Robert (ed.) *New Dimensions,* vol 3. Nelson Doubleday/SFBC.

Losi, M. E., Amrhein, C., & Frankenberger, W. T. (1994). Environmental biochemistry of chromium. *Reviews of Environmental Contamination and Toxicology,* 91–121.

Lottermoser, B. (2007). *Mine wastes characterization, treatment, environmental impacts* (2nd ed.). Springer.

Lu, Q., Xiaohang, X., Liang, L., Zhidong, X., Shang, L., Guo, J., Xiao, D., & Qiu, G. (2019). Barium concentration, phytoavailability, and risk assessment in soil-rice systems from an active barium mining region. *Applied Geochemistry, 106*, 142–148.

Manhart, A., Blepp, M., Fischer, C., Graulich, K., Prakash, S., Priess, R., Schleicher, T., & Tür, M. (2016). *Resource efficiency in the ICT sector.* Greenpeace, Oeko-Institut eV.

Maxwell, R., & Miller, T. (2012). *Greening the media.* Oxford University Press.

Merchant, B. (2017). *The one device: The secret history of the iPhone.* Hachette.

Monteiro, W. A. (2014). The influence of alloy element on magnesium for electronic devices applications–a review. *Light Metal Alloys Applications, 12*, 229.

Mudd, G. M. (2007). Global trends in gold mining: Towards quantifying environmental and resource sustainability. *Resources Policy, 32*(1–2), 42–56.

Mudd, G. M. (2010). Global trends and environmental issues in nickel mining: Sulfides versus laterites. *Ore Geology Reviews, 38*(1–2), 9–26.

Nest, M. (2011). *Coltan*. Polity.

New Jersey Dept of Health and Senior Services. (2001). *Hazardous substance fact sheet: Gallium*. https://nj.gov/health/eoh/rtkweb/documents/fs/0956.pdf. Accessed 1 Apr 2021.

Nickel Institute, The. (2021). *Stainless steel: The role of nickel*. https://nickelinstitute.org/about-nickel/stainless-steel/. Accessed 1 Apr 2021.

Norgate, T. E., Jahanshahi, S., & Rankin, W. J. (2007). Assessing the environmental impact of metal production processes. *Journal of Cleaner Production, 15*(8–9), 838–848.

Oguchi, M., Murakami, S., Sakanakura, H., Kida, A., & Kameya, T. (2011). A preliminary categorization of end-of-life electrical and electronic equipment as secondary metal resources. *Waste Management, 31*(9–10), 2150–2160.

Olsgard, F., & Hasle, J. R. (1993). Impact of waste from titanium mining on benthic fauna. *Journal of Experimental Marine Biology and Ecology, 172*(1–2), 185–213.

Paraskevas, D., Kellens, K., Van de Voorde, A., Dewulf, W., & Duflou, J. R. (2016). Environmental impact analysis of primary aluminium production at country level. *Procedia CIRP, 40*, 209–213.

Periodic Videos. (2013, May 28). Super expensive metals – Periodic table of videos. *YouTube*. https://www.youtube.com/watch?v=Fg2WzCzKpYU. Accessed 2 Apr 2021.

Pradhan, D., Panda, S., & Sukla, L. B. (2018). Recent advances in indium metallurgy: A review. *Mineral Processing and Extractive Metallurgy Review, 39*(3), 167–180.

Pukas, A. (2016, January 30). What became of the Chilean miners five years on? *Express*. https://www.express.co.uk/news/world/639433/Chilean-miners-the-33-antontonio-banderas-juan-illanes-San-Jose-mine. Accessed 1 Apr 2021.

Rao, M. N., Sultana, R., & Kota, S. H. (2017). *Solid and Hazardous Waste Management*. Butterworth-Heinemann. https://doi.org/10.1016/B978-0-1 2-809734-2.00006-7

ROHS Guide. (2021, April 2). *2021 ROHS compliance guide. Regulations, 10 substances, exemptions*. https://www.rohsguide.com/. Accessed 2 Apr 2021.

Schulz, K. J., DeYoung, J. H., Seal, R. R., & Bradley, D. C. (Eds.). (2017). *Critical mineral resources of the United States: Economic and environmental geology and prospects for future supply*. US Geological Survey.

Shanker, A. K., Cervantes, C., Loza-Tavera, H., & Avudainayagam, S. (2005). Chromium toxicity in plants. *Environment International, 31*(5), 739–753.

Taffel, S. (2012). Escaping attention: Digital media hardware, materiality and ecological cost. *Culture Machine, 13*.

Taffel, S. (2015). Towards an ethical electronics? Ecologies of Congolese conflict minerals. *Westminster Papers in Communication and Culture, 10*(1).

Takahashi, K., Sasaki, A., Dodbiba, G., Sadaki, J., Sato, N., & Fujita, T. (2009). Recovering indium from the liquid crystal display of discarded cellular phones by means of chloride-induced vaporization at relatively low temperature. *Metallurgical and Materials Transactions A, 40*(4), 891–900.

T.E., Norgate S., Jahanshahi W.J., Rankin (2007). Assessing the environmental impact of metal production processes. *Journal of Cleaner Production* 15(8–9) 838-848 10.1016/j.jclepro.2006.06.018

U.S. Geological Survey. 2014. *Platinum group metals: So many useful properties.*

U.S. Geological Survey. (2018). *2016 Minerals Yearbook.* https://www.usgs.gov/centers/nmic/minerals-yearbook-metals-and-minerals

U.S. Geological Survey. (2019). *Mineral commodity summaries 2019: U.S. Geological Survey.* https://doi.org/10.3133/70202434.

U.S. Geological Survey. (2021a). *Nickel statistics and information.* https://www.usgs.gov/centers/nmic/nickel-statistics-and-information. Accessed 1 Apr 2021.

U.S. Geological Survey. (2021b). *Copper statistics and information.* https://www.usgs.gov/centers/nmic/copper-statistics-and-information. Accessed 1 Apr 2021.

USGS. (2016). *2016 Minerals Yearbook [Advance Release].* U.S. Department of the Interior, U.S. Geological Survey.

USGS. (2017). "Barite (Barium)". *Chapter D in Critical Mineral Resources of the United States—Economic and Environmental Geology and Prospects for Future Supply.* Professional Paper 1802–D. U.S. Department of the Interior, U.S. Geological Survey.

Wang, L., Tai, P., Jia, C., Li, X., Li, P., & Xiong, X. (2015). Magnesium contamination in soil at a magnesite mining region of Liaoning Province, China. *Bulletin of Environmental Contamination and Toxicology, 95*(1), 90–96.

Wilburn, D. R. (2015, September 22). *Lead scrap use and trade patterns in the United States, 1995–2012.* US Geological Survey. https://pubs.er.usgs.gov/publication/sir20155114

World Health Organization. (2017, March 31). *Mercury and health.* https://www.who.int/news-room/fact-sheets/detail/mercury-and-health. Accessed 2 Apr 2021.

Yang, C., Tan, Q., Zeng, X., Zhang, Y., Wang, Z., & Li, J. (2018). Measuring the sustainability of tin in China. *Science of the Total Environment, 635,* 1351–1359.

Yusof, A. M., Mahat, M. N., Omar, N., & Wood, A. K. H. (2001). Water quality studies in an aquatic environment of disused tin-mining pools and in drinking water. *Ecological Engineering, 16*(3), 405–414.

# CHAPTER 8

# Where to from Here?

Throughout this book I have tried to show that a carbon neutral games industry is not just desirable, but that it is both possible and absolutely essential. I have tried to demonstrate that decarbonization must be the top priority, in part by making clear just how inappropriate and inadequate it is to rely on games powers of persuasion of others while remaining dependent on fossil fuels to produce. This is accentuated when it comes to controversial or ideological topics, where it is so much more effective to simply change the world for the better in those places where we already have the power to do so. The games industry has a moral responsibility to change itself before it tries to change others, taking the proverbial plank out of its own eye before it reaches for the spec in players'. What this means is simple in theory and difficult in practice: reducing and eliminating carbon emissions and practicing careful environmental stewardship. This is utterly essential, a pre-requisite to the truly ecological game. When even 'green' games can be made with power provided by fossil fuels, sold to players on plastic media that is guaranteed to end up in landfill, made for and played on devices that are some of the most resource intensive objects the planet has ever seen, there is clearly something deeply wrong with our priorities, and our understanding of what counts as 'green'. It is not enough, cannot seriously be considered ecological, if with one hand

B. J. Abraham, *Digital Games After Climate Change*, Palgrave
Studies in Media and Environmental Communication,
https://doi.org/10.1007/978-3-030-91705-0_8

games promote sustainability and positive environmental concern while with the other rely on the same energy and resource intensive patterns, the same fossil fuel driven systems and extractive logics of capitalism for its own existence.

I have tried to show that are ways to change, and to reduce emissions and other environmental harms from games. But we also saw just how substantial those emissions are. Figures like the estimated 15 million tonnes of $CO^2$ equivalent emissions generated by the process of making games in a single year, ballooning even higher if we count travel. We considered the emissions of sending discs to every corner of the globe—just the emissions from shipping coming out to as much as 800 tonnes per annum to one relatively minor market (Australia). We also saw Aslan's (2020) figure of as much as 27 TWh of power consumed, by a the PS4 console in Europe over a few years—perhaps 18 million tonnes of $CO^2$ emissions. Mills et al.'s (2019) estimate for the United States lead to a figure of around 24 million tonnes of $CO^2$ equivalent emissions just for one year. Finally, I tried to show just how tenacious the deeply embedded emissions and other environmental harms are within gaming hardware itself, tracing some of the atomic elements present in a common (though now dated) gaming console back to their source in the earth. Taken together, these chapters and the picture they paint represent a serious—though perhaps not utterly damning—portrait of the modern games industry and its energy and resource intensity. There is so much work still to be done, and a great deal of room for future researchers to take up this challenge of both more accurate measurement and estimates, and better understanding of options for reductions.

The games industry must lead by example, and begin to change to meet the demands of the climate crisis head on, in ways both big and small. We no longer have the luxury of tinkering around the edges—only deep and lasting structural changes, collective and individual actions on a scale that seems quite daunting to many, will be sufficient to meet the challenges ahead. Knowing our desired end-point—the carbon neutral games industry—is also not the same as knowing the path to achieve it, and a secondary goal of this book has been to identify the contours, the barriers and challenges along that path, pointing out some of the ways we might get from here to there. From Chap. 4 onwards, the order of chapters followed the path from where I see the easiest most immediate and direct actions to be taken, ending where I suspect full decarbonization will remain most difficult to achieve. Yet all of these emissions I outlined here will need addressing—some through an almost total reconfiguration of the industry, even ultimately the complete abandonment or transformation of the system of capitalist development that we have been saddled with for so long. Some changes have already been proven possible, can already be done quickly and within existing market-economic frameworks. Some of the problems outlined here, like corporate control over closed platforms that makes digital distribution less appealing, are not so much a 'bug' as a 'feature' of modern capitalism, requiring an organized struggle to put a cap on. Above all, the unlimited, unending growth of the past is simply not compatible with the preservation of life and biodiversity on earth as it fuels the dangerous warming that threatens everything, sooner or later. I take great encouragement from seeing more and more ordinary people coming to the same conclusion—from school student strikers, to workers for climate action, coalitions of scientists and experts and ordinary concerned citizens. A profound rethinking of our political-economic systems are needed, just as much as more prosaic actions like better recycling of our e-waste. Such a transformational project is beyond the ability of this book to catalyse, perhaps even beyond the realms of imagination for many readers. If nothing else, this book has lifted the lid on the substantial carbon emissions that the games industry is responsible for, what it does in response to this knowledge now lies in its hands. From industry leaders, to game developers, game journalists, researchers, activists, players and more both inside and outside the industry, as well as national and regional policy makers. There are dozens of roles to be played in the increasingly urgent work of decarbonization, even within the games industry.

In this final chapter I want to focus on what can be done in the immediate future, and what the next steps are in the pathway to a carbon neutral games industry. We can no longer ignore the emissions of the digital games sector—whether they occur in the home office, the game development studio teeming with hundreds, whether they emerge from manufacturing and logistical hubs shipping discs and consoles around the world, or from the server racks hosting and transmitting files and game data, even in the very mining pits that dig up the raw materials that go into each gaming console—there are harms to the planet, and ultimately to ourselves, from each.

What is to be done? Here I get more explicit about the kinds of actions I think are warranted and likely necessary in getting to a carbon neutral games industry. I have split the following recommendations and potential actions to take into categories for different groups: actions that can be taken by game developers, actions that can be taken by hardware manufacturers/platform holders, and those that can be taken by game players, journalists, and others in and around the games industry's periphery. Again, these are not meant to be exhaustive or prescriptive, but generative of further ideas and discussion. They are intended to be indicative of the broad types of action that will be needed to get us to the carbon neutral games industry.

## Action for Game Developers

Starting with game developers, as they have perhaps the most to do immediately, and are best placed for a variety of fast actions to decarbonize the game industry in the short and long term. Game developers have some control over the kinds of games that get made, but perhaps more importantly, they can have control over the way that they make those games. A tricky question for all these suggestions is how this might actually be achieved, with so many developers far removed from the actual levers of power and decision-making in their own workplace. As a result, my first suggestion is a way to force change in the industry.

First, I suggest forming climate councils of concerned developers, and finding in your workplace others who care about the same issues. Together you can discuss actions and raise visibility of emissions as a group. Where developers face receptive and accommodating leadership (such as we saw in the Wargaming.net discussion in Chap. 4) recommendations can be taken to leaders for swift implementation. For non-receptive or indifferent bosses and owners, having a group that can advocate collectively and agitate for change will be crucial, as building and exerting collective power among workers is a tool that has been used effectively for hundreds of years. Environmental and climate issues are increasingly recognized as central to the work unions do, and many of the following actions will only be possible with the buy-in and solidarity that is possible through a unionized workplace with colleagues working towards the same goals, able to put pressure on decisionmakers and influence the direction of the games industry itself. Unionizing the games industry is not without its challenges (see for instance discussions by Weststar and Legault (2017), Abraham and Keogh (2021), Ruffino and Woodcock (2021)), however, it may prove essential in getting the owners of a studio, a publisher or other big corporation to listen.

Whether together or alone, in the wide variety of places and settings that game development happens, there are many other actions that game developers can take:

- Collect and publish workplace energy consumption and emissions, including business travel, waste sent to landfill, etc. and set goals for reductions.
- Switch your studio or office electricity to 100% renewable or carbon offset power—we saw in Chap. 4 the electricity used by game development studios is a substantial cause of emissions, and one of the quickest and easiest places to decarbonize, sending important signals to the market about fossil fuel generation being unacceptable. It is also a pre-requisite to the truly ecological game, with Scope 1 and 2 emissions sole responsibility of developers.
- Collect and publish play–duration data for your games, broken down by country or region so it is possible to estimate players' Scope 3 emissions accurately.
- Advocate for and where economical purchase carbon offsets for Scope 3 emissions from players until more effective carbon neutral solutions emerge, such as fully renewable power systems. Ensure the credibility of these offsets by making sure they are certified to at least the UN standard (see https://offset.climateneutralnow.org/UNcertification).
- Where purchasing offsets is not economically feasible, building in a carbon emissions price into the cost of future games and products may be an option instead, particularly for large organizations.
- Sell games digitally whenever possible—avoid physically shipping plastic discs and cartridges around the world, as these emissions are difficult to reduce and the physical media itself will almost inevitably end up in landfill. Where you must have physical releases, investigate the use of bioplastics and cardboard alternatives, and other substitutions made from biodegradable or fully compostable materials.
- Make a substantial impact on embodied emissions in devices by doing what you can to extend the lifespan of the hardware you use—servers, workstations, monitors, etc.—and the hardware that you support. Making your players and your own devices last longer will contribute to discouraging unnecessary upgrades. Many of these devices are still not fully recyclable, wasting their precious materials. (See Chap. 7)

- Once the essential reduction of development and distribution emissions is complete, work on reducing the power demands made on player's hardware through changes to the games you make. Reducing Scope 3 emissions of players is possible (see Chap. 6), however it will likely take a lot of work for uncertain reward, at least compared to some of the previous actions. Scope 3 emissions are also more easily tackled via greater penetration of renewable power generation, which game developers can support with their own power purchasing.
- Tell the world and the rest of the industry what you are doing—celebrate your successes and make noise about the industry's need to change. Many game developers are already starting the journey of reducing and eliminating their own emissions, but even industry leaders could be much more vocal about their achievements (see Chap. 4).

Taken together, these are just the first few steps on the long road to carbon neutrality, but even these are poised to have a big impact on reducing carbon emissions and speeding the transition along. This will be an ongoing process as momentum gathers and more developers, studios, and corporations get on board, and as our understanding of both the crisis and what it demands from us improve. As long as we keep carbon neutrality as our goal, and take some action towards it as quickly as we can, we can probably still keep much of what we love about the games industry today. There will be changes, and hard decisions to make, especially towards the "end" of this generation of console's lifespan, and at other difficult points. Some of these changes will present new problems and challenges, but with new challenges come new opportunities, ones that might let us rethink what the games industry looks like, how it operates, who it is for, even what we are willing to give up or let go of in order to keep our planet a safe and sustainable place to live for generations to come.

## Hardware Manufacturers, Publishers and Platform Holders

Despite potentially perhaps the most difficult to fully decarbonize (see Chap. 7) with precious few alternatives to digging materials out of the ground, the hardware and manufacturing side of the games industry plays a significant role in determining the long-term base line energy demand of the modern games industry. Consoles and PC gaming hardware determine the present and the future of the industry, from setting clock speeds and memory capacities with the design of new consoles, to determining which games are allowed to be published on a particular platform. Physical devices represent substantial embodied emissions through manufacturing, and induce power demand through consumers hungry for the latest and greatest gadgets. There are many things the likes of Sony, Microsoft, Nintendo, and others can do to contribute to a carbon neutral games industry. Some broad goals and actions for hardware makers, publishers and other platform holders include:

- Extending console lifecycles, which in turn reduces the emissions generated by manufacturing new hardware. This could be done through simply not making new hardware, by changing the way these devices are made, increasing the scope for upgrades, user repairs, and better expandability.
- Making devices more easily repairable can help avoid waste, landfill and lower embedded emissions. This applies to all hardware, from consoles to controllers, peripherals and other devices.
- Audit supply chains to ensure suppliers are providing sustainable materials and sourcing them responsibly. There are already regulatory requirements around many conflict resources, but extending this care in procurement to all materials will be essential. Likewise hardware owners can put pressure on manufacturing partners to disclose emissions from assembly, shipping, etc.
- Increased efforts to improve energy efficiency in hardware revisions can result in bringing down end consumer energy demands. (See Chap. 6) The state of California has already implemented regulations preventing computing hardware energy consumption above a certain level—surely a sign of things to come. Manufacturers may be well served by paying attention to this trend, as it presents both an opportunity and a threat to their business.
- Consider ways to make digital-only purchasing of games more attractive, more equitable, and more widespread. Avoid emissions from transporting discs and the stream of plastic waste ending up in landfill.
- Accurate carbon accounting and disclosure, even mandating carbon neutrality, could become a condition of funding for games by publishers, or a requirement for publishing on certain platforms. Prioritize carbon neutral data centers when making this switch.

Some of these proposed changes will have consequences for the bottom line, and perhaps they are unlikely to be made willingly, except by the most forward thinking of leaders. External pressure, new regulations, activist shareholders, even well-organized consumer boycotts may further tip the scales towards change. Significant pressure may be brought to bear through the application of collective power—both from unionized game developers and game journalists, as well as through the power of consumer action, boycotts, divestment and sanctions. Some of these may end up being more symbolic than anything else, but there may still be a benefit to these symbolic gestures. Again, the larger forces at play here are the capitalist system and the modern publicly traded corporation, both of which present real impediments to change, as they are deeply invested (both figuratively and literally) in the continuation of the status quo. We should not be put off or disheartened by the challenge ahead of us, but instead take courage from the fact that one way or another, change is coming.

## Players and Everyone Else

Lastly, for game players and others in the peripheral orbit of the industry there are actions to take as well. This includes those who make a living from streaming games, journalists, youtubers and other content creators, as well as retail workers and the average game player. There is plenty to do for those of us not directly involved in making games:

- Purchase renewable power for our own homes and offices to cover the energy demands of our gaming. Our electricity consumption is our responsibility, and one of the few places we have control over our own emissions.
- Use the power saving options on consoles and device, turning them off when not in use. This has the added benefit of reducing your power bills and saving money.
- Repair devices where possible, and pursue incremental upgrades to extend the lifespan of a device rather than replacing it—this is not always possible because of the way devices are designed, but websites like iFixit.com often have guides for fixing things like controllers when issue arise. There is a deep satisfaction to be had in fixing one's own devices.

- Buy games digitally wherever possible, and if you have to buy discs try and buy second hand.
- Streamers can take note of and share the energy used streaming, reducing power consumption also helps in reducing power bills. Normalize talking about emissions and that you are doing something about them with your audience.
- Journalists can ask developers whether their games are carbon neutral, what plans they have to achieve it, and when they aim to achieve it by, and include carbon emissions in part of the review process and scoring. Bring climate perspectives into your reporting on all new games, game hardware and peripherals.
- Everyone, including players themselves, can ask the makers of the games they play where they stand on climate issues, and what they are doing about it. Game companies are keenly aware of their audience's expectations, and letting them know you're paying attention to their actions on emissions will only help.
- Join a climate activist or advocacy group near you—the Climate Action Network has a list of organizations in over 130 countries (see: https://climatenetwork.org)—to collectively advocate for systemic change and the transition to renewables.

The list of actions for consumers and those of us on the periphery of the industry is sadly, but perhaps unsurprisingly, limited, reflecting the relative powerlessness of individuals in the face of global corporations and the problem of climate change. Forming collectives, shifting norms, and building climate awareness in and around the industry is still necessary, and will probably require us to pool resources and build power through broad coalitions. Different groups around the games industry have different tools and leverage at their disposal, and we should use whatever we have. Consumer actions only have power when taken en masse, and organizing things like boycotts and awareness campaigns targeting carbon intense companies can be prohibitively difficult for even the biggest, most well-funded and organized activist groups. They can also syphon energy from more important work like political pressure and lobbying, and direct action. We can only do so much, but we shouldn't let the limitations of the present stop us from imagining the potential of the future. We can acknowledge our limits and still find ways to practically support each other and those who may be better placed to act.

## FIGHTING FOR A GREEN FUTURE FOR THE GAMES INDUSTRY

It is difficult to think of a time where the entire games industry has come together to act in unison. More typically, it is dominated by competition, platform wars, and fights over market and mind share—but perhaps it is time for change. One issue that plagued much of the early negotiations over emissions reductions, from the early 90s until the signing of the Paris agreement in 2015, were arguments over who should be the first to act. Just like how certain countries saw emission reductions not through the lens of a moral imperative but through the lens of the competitive disadvantage, some parts of the games industry may be tempted to resist calls for emissions reductions and a carbon neutral target. Thankfully, the challenge of being the first to act has already been overcome—demonstrated by a number of developers as we saw in Chap. 4. It may not be the case that there are *no* disadvantages to acting on carbon neutrality, but there is plenty of proof they can be overcome. This needs to continue until the tide turns and the pressure becomes overwhelming, with fossil fuels already close to tipping points of unprofitability. There is still no enforcement or incentive mechanism in the games industry however. The UN *Playing for the Planet* initiative is still one of only a few industry bodies visibly concerned with the climate impact of the games industry, and its membership is entirely voluntary. Even so it is somewhat disappointing to see that not even all members of this group have committed to carbon neutrality or set concrete deadlines. The 2020 *Playing for the Planet* progress report includes updates from 19 member organizations, with less than half having committed to full carbon neutrality (UNEP, 2020). Those that have are to be applauded and monitored to make sure they meet their commitments. But if even those nominally committed can be behind the curve, how much more the bulk of the games industry? There is a long way to go.

We also know that carbon neutral targets, while essential, are not on their own sufficient—it makes a big difference how quickly we achieve carbon neutrality and how much is emitted in the interim. Nominally, there may be hope that national carbon inventories can play a part on keeping the earth within the 1.5 °C "safe" warming limit, and policymakers may play a part in designing incentives and smoothing the path of this process, though I too have my reservations about the sufficiency of carbon taxes and other mechanisms, following Mann and Wainwright's (2018) criticisms in *Climate Leviathan*, and Bryant's (2019) analysis in *Carbon*

*Markets in a Climate-Changing Capitalism.* Many of the world's countries now have 2050 (or sooner) targets for carbon neutrality, and game companies situated in these regions of the world will be well served by getting in early and avoiding the disincentive 'stick' of tariffs, taxes and other mechanisms. Facing a relatively straightforward transition compared to some industries, getting in ahead of the curve may allow organizations to avoid tougher interventions. The games industry has enormous potential to take action already, but it will need to do better than it has so far.

I conclude by challenging those figures in the games industry who consider themselves industry leaders—CEOs, lead designers, creative directors, journalists with audiences of tens or hundreds of thousands—who are not yet fully committed to carbon neutrality in their own companies and organizations to get on board or get out of the way. Climate change is coming, but so are we.

## REFERENCES

Abraham, B., & Keogh, B. (Forthcoming 2022). Challenges and opportunities for collective action and unionization in local game industries. *Organization.*

Aslan, J. (2020). *Climate change implications of gaming products and services.* PhD dissertation, University of Surrey.

Bryant, G. (2019). *Carbon markets in a climate-changing capitalism.* Cambridge University Press.

Mills, E., Bourassa, N., Rainer, L., Mai, J., Shehabi, A., & Mills, N. (2019). Toward greener gaming: Estimating national energy use and energy efficiency potential. *The Computer Games Journal, 8*(3), 157–178.

Ruffino, P., & Woodcock, J. (2021). Game workers and the empire: Unionisation in the UK video game industry. *Games and Culture, 16*(3), 317–328.

UN Environmental Program. (2020). *Playing for the planet annual impact report 2020.* https://playing4theplanet.org/2020annual_impact_report/

Wainwright, J., & Mann, G. (2018). *Climate Leviathan: A political theory of our planetary future.* Verso Books.

Weststar, J., & Legault, M.-J. (2017). Why might a videogame developer join a union? *Labor Studies Journal, 42*(4), 295–321.

# INDEX